A Strategic Approach to the UK Construction Industry

The construction industry is a microcosm of the economy as a whole, and as such the economics of the sector contains many of the aspects of the economy in general, albeit with fascinating and unique features. What are the implications of economic theory for the future of UK construction? How does the industry ensure innovation, quality and efficiency? What priorities might best serve the construction industry, those working in it and their customers?

In seeking answers to these and other questions, the UK government has commissioned a number of reports on the construction industry including the Latham and Egan Reports and more recently *Construction 2025*. These have invariably proposed time and cost targets for the construction industry. In this new book, Stephen Gruneberg stands in stark contrast to those reports and presents the relevant theoretical aspects of construction economics to account for the behaviour of construction firms and suggest a strategy for future growth and sustainability. He discusses the theory and data relating to the output of firms in relation to the type of firm, the market and how these firms behave as a result. The purpose of this book is to advocate for the measures needed to create the kind of industry that must be fostered to ensure the quality of its output, sustainability and fair terms and conditions of employment for its workforce.

Gruneberg's new book is essential reading for anyone wishing to understand the economic forces that determine industry outcomes and who has a stake in the success of the UK construction sector.

Stephen Gruneberg is an Honorary Professor at the Bartlett School of Construction and Project Management, UCL.

A Strategic Approach to the UK Construction Industry

Stephen Gruneberg

Routledge
Taylor & Francis Group

LONDON AND NEW YORK

First published 2019
by Routledge
2 Park Square, Milton Park, Abingdon, Oxon OX14 4RN

and by Routledge
52 Vanderbilt Avenue, New York, NY 10017

Routledge is an imprint of the Taylor & Francis Group, an informa business

British Library Cataloguing-in-Publication Data
A catalogue record for this book is available from the British Library

Library of Congress Cataloging-in-Publication Data
Names: Gruneberg, Stephen L., author.
Title: A strategic approach to the UK construction industry / Stephen Gruneberg.
Description: Abingdon, Oxon ; New York, NY : Routledge, 2019.
Identifiers: LCCN 2018046053 | ISBN 9781138089778 (hardback) | ISBN
    9781315109022 (ebook)
Subjects: LCSH: Construction industry—Great Britain.
Classification: LCC HD9715.G72 G78 2019 | DDC 338.4/76240941—dc23
LC record available at https://lccn.loc.gov/2018046053

ISBN: 978-1-138-08977-8 (hbk)
ISBN: 978-1-315-10902-2 (ebk)

Typeset in Times New Roman
by Apex CoVantage, LLC

Contents

Preface

This book has been written in response to the many official reports on the construction industry mainly since the Second World War. In that time one predominant theme has been the use of targets to modernise the construction industry. This short book attempts to steer away from that path towards considering the priorities that might be used to define the aims and purpose of an aspirational construction sector.

The book begins with an overview of reports on the construction industry since the Simon Report of 1944 to the latest Construction Sector Deal published by the Department for Business, Energy and Industrial Strategy in 2018. In Chapter 2, I discuss very low profit margins preventing firms from investing in new methods, employing labour directly or cutting down on training. Nevertheless, far from being a backward industry, the construction sector, almost unnoticed, continually innovates and improves.

The challenge faced by firms in the construction industry is that the profit margin of contractors is very low and their turnover is very high. This accounts for the behaviour of firms in construction – casualisation of labour, cost cutting and lack of trust as well as delaying payments to suppliers.

Chapter 3 argues that the policy of using industry-wide cost, time and environmental targets is ineffective and should be replaced with objectives that relate to the building industry and the built environment itself.

These priorities are set out in Chapter 4 and include the need for the construction industry to be both competitive and produce a high quality output, with a professional workforce, using efficient production processes. If change is to come to the construction industry there is a need to define it, to say what is included and excluded and what the barriers to change may be. In Chapter 5 there is a discussion about the size and scope of construction. Chapter 6 deals with productivity and innovation in construction and defends construction firms from the accusation that delays are invariably caused by contractors.

Chapter 7 deals with the need to be internationally competitive to give reassurance that the performance and standards of the industry are internationally comparable. Although Chapter 8 is entitled *Efficiency and Professionalism*, it is really concerned with the need to change the culture in the construction industry. Chapter 9 reinforces the cultural changes needed within construction by arguing that a new industry-wide body is needed to integrate thinking and attitudes in construction. A new professionalism in construction would also then be promoted. To complement these supply side aspects of construction, the following chapter discusses the need to strengthen the demand side by ensuring a steady flow of orders. This will enable contractors to have confidence in their own long run survival and take the necessary measures. These measures are discussed in Chapter 10.

Government intervention is seen as necessary as the industry is not capable of reforming and controlling itself. The reason is that it is such a fragmented industry that without outside intervention and a clear set of rules, poor practice and undercutting will tend to undermine the health and productive potential of the network of firms that make up the construction sector. In the final chapter, Chapter 11, is a discussion about a proposed new Ministry of Construction to oversee and co-ordinate this important sector of the economy.

I am very grateful to all at Taylor & Francis for the help they have given me in producing this volume, including Ed Needles, Liz Spicer and Patrick Hetherington, and Jennifer Bonnar and Autumn Spalding at Apex CoVantage. Needless to say, all the errors are entirely down to me!

Stephen Gruneberg

1 Introduction

Tracing the history of construction reports from Simon (1944) to the Construction Sector Deal (2018)

Introduction

Due to the high proportion of very small building contractors, the construction industry is incapable of carrying out certain functions, such as training, without the kind of rules that governed the activities of guilds in the distant past or without some form of compulsory grant levy system, where appropriate. A grant levy system has been operated by the Construction Industry Training Board (CITB) since the 1960s. All contractors must pay a levy depending on the size of the firm. This fund is then used to distribute grants to those firms, which undertake training and employ apprentices. Direct government intervention is therefore essential if the public sector is to intervene directly in the funding of training needs in the construction industry. Otherwise, the industry itself does not have the means or structure to meet the priorities and needs of this or any construction policy or objective. For example, if the workforce is incapable of meeting the skill requirements of technologically advanced techniques and new environmental demands, additional help may be required.

In spite of these weaknesses, the construction industry is very responsive to changes in demand – for example, when major projects are put to the industry, it is possible to mobilise resources, such as plant and equipment, labour and even materials, at least sufficient to prepare a site for development, within a very short space of time. Firms in the construction industry also continually adapt to the challenge of new materials and technology and remain an innovating sector of the economy as new products and materials are introduced on site and integrated into the construction process. However, many traditional practices appear to be embedded in employment practices of the past and influence the way the industry operates. Procurement practices, the role of the architect, the planning system and land ownership are all rooted in the traditional past. The culture of the industry itself, as a whole, tends to reinforce the traditional fragmented nature of the sector.

Construction is a project orientated industry. As a result, everything to do with construction management is of a temporary nature. People come together only to work on a project, and on completion they disperse to work on many other jobs. Although many in the industry recognise the advantages of keeping project teams together to work on different projects, this very rarely occurs in construction. As a result, construction teams tend to be temporary organisations with an impaired ability to create organisational knowledge that is cumulative and that is common practice in many other sectors of the economy. It is rarely possible to know the strengths and weaknesses, personalities and preferences of individual staff working on site. This means that the allocation of work to individuals is often sub-optimal, and people can be asked to carry out work they are not equipped to carry out or capable of doing. In spite of these shortcomings, new methods of working and new materials are constantly being introduced on site.

Any industrial strategy for the construction industry must take into account the nature of the industry and how it operates. Before proposing a remedy it is important to understand the causes of the problems in the construction industry in the first place, and there is invariably more than one cause.

The behaviour of individual contractors and other firms is to some extent determined by the specific economic features of the construction sector. These features constrain the different options available to firms in the construction industry as a whole and account for many of the problems beyond their control that face contractors and their clients. For example, construction firms work on very small profit margins leaving very little room for error or room for negotiation, especially when things go wrong on site. Their small profit margins force firms to expand turnover to such an extent that the level of risk and default is greater in construction than in almost any other sector of the economy. One would expect that the larger the project the greater the profit margin, but this is not necessarily the case in construction. Many of the problems of construction arise out of the high risk faced by contractors, including the risk of losing money on a project where profit margins can be easily eroded altogether. As a consequence it is not surprising to find that the construction industry is litigious and confrontational. The very low profit margins on a project mean there is no room for negotiation between firms working on projects. This matters as the construction industry is a major component of the economy.

The size and significance of the construction industry compared to the national economy are far greater than the figures given in official data imply. This is because official data is based only on the value added on site by main and specialist contractors and not on the value of the output of construction in terms of the finished goods it produces or the value over many years of the services provided by buildings and infrastructure once completed.

The conventional measurement of the size of the construction industry is based only on the value added to inputs by contractors on site. While this approach is entirely appropriate for finding the aggregate size of the economy, because the value added approach avoids double counting, it does not take into account the sheer size and scale of building projects and the impact they have on the economy. The conventional method of calculating the relative importance of construction therefore underestimates the size of the construction sector's contribution to the economy, which is significantly greater when building components and materials, design and civil engineering consultants and other construction inputs are included. The significance of the building industry can be better assessed as the annual value of the *final* built goods, not just the value added on site by contractors. In assessing the size of construction one intuitively includes the materials that go into the production of the built environment. This more intuitive measure of the significance of the production of the built environment as a percentage of GDP includes the value added by contractors as well as the value of steel, concrete and glass and all the other material and design inputs used in the production process, equivalent to 15.3 per cent of the whole economy, according to Table 5.1 (see Chapter 5).

There has been no shortage of reports on the health of the construction industry and prescriptions for its recovery over the past 30 years or more. The most comprehensive list of reports on the UK construction industry is given on the website of Designing Buildings Wiki (2017) with no fewer than 79 reports named (see the Appendix for the list of reports from 1934 to 2018). Nor are the phenomena found in the UK unique to the UK building industry. King (1996) found much in common between the US and the UK.

In spite of all this effort little has changed in terms of the criticism and practice of construction in the UK over the past 50 years. The same comments keep re-occurring: late delivery, inadequate quality and poor budgetary control. In spite of efforts to the contrary, the industry seems unable to improve. Perhaps there are good reasons for this disappointing state of affairs. We begin by looking at what others have said, by referring to many of the reports that have highlighted the problems.

The Simon Report

Even before the Second World War had drawn to an end, the state of the construction industry had been a cause for concern. Would the construction industry be able to cope with the increased need to build housing after the war? Many homes had been destroyed. Others were old and dilapidated, and returning soldiers would need housing. In 1944 a committee under the chairmanship of Sir Ernest Simon was commissioned to prepare a report on the construction industry entitled "Placing and Management of Building Contracts."

One of the themes of the Simon Report was procurement and the contractual arrangements in place to facilitate speedy construction. Site management was also an issue due to the volatility of the market, which made it difficult to employ labour. Even at a time of rationing, there was criticism that trade associations were fixing prices, and the Simon Report called for all price fixing agreements to be registered.

The Emmerson Report

In the 70-odd years since the Second World War there have been at least 77 government reports on the construction industry. The Emmerson Report of 1962 was briefed to examine the barriers to improving efficiency. The report emphasised the role of architects but found them wanting in terms of cost control and managing the building process, which had by 1962 become a complex set of relationships between main contractors and several sub-contractors. Nevertheless, Moodley and Preece (2003) point out that the Emmerson Report highlighted the gap between architects and builders, a gap that was greater than similar relations between designers and producers in any other industry.

Emmerson also discussed the issue of training of both operatives and management, including the length of apprenticeships and the ability of apprentices to move between firms. The Emmerson Report recognised the importance of regular and continuous employment. Otherwise attracting and retaining staff would be problematic. Emmerson also saw a possible role for government departments in their relations with firms to encourage construction firms to be more efficient.

Although the contracting system had been in existence since Victorian times and was in urgent need of revision, Emmerson noted that proposed reforms put forward in 1954 by the Robertson Committee on selective tendering and fixed price contracts had not been adopted. Instead open competitive tendering meant that too many firms were competing for work, and the winning lowest tendered price was often not profitable for the winning contractor, leading to disputes and problems for all. Economists call this the winner's curse. As Banwell (1964) was to reiterate in his report, winners in open competitive tender competitions frequently had neither the skills nor the capacity to undertake the work. This failure to modernise the industry to meet modern demands did not appear to have been taken on board by clients or contractors.

One feature that seems to apply to all government reports on the building sector is the assumption that the reports might actually in some way influence change in the construction industry. Reports alone do not necessarily change anything. This may appear to be a somewhat cynical point of view, but it is not clear what the mechanism is for executing the changes recommended. A military model comes to mind, namely the government commissions a report, debates its recommendations and hands over the agreed proposals to a general who then orders others lower in the hierarchy to carry out the changes needed to meet the aspirations of the report.

Instead, there is a highly fragmented structure of independent firms and organisations, each pursuing its own interests, namely survival and the making of profits. The troops, either the small specialist firms or even the individual skilled and unskilled workers, are rarely involved in the strategic discussions regarding the future of their industry. It is hardly surprising that change in the construction industry is slow or static. The business model does not fit the implicit assumption made in these reports, which is that the self-interest of the major firms and professional bodies that participate in the research and writing of the reports is sufficient to bring about reform of the industry. For example, it is not surprising that, according to Moodley and Preece (2003), Emmerson noted in 1962 that measures that had been recommended in the 1950s had simply not been adopted.

The Banwell Report

Only two years after the Emmerson Report the Banwell Report was published in 1964. Its title was "The Placing and Management of Contracts for Building and Civil Engineering Work." Its focus was current practice regarding contracts and improvement of performance. Inadequate time to prepare in advance of actual construction was seen to lead to problems on site causing additional expenses and delays. To solve the problem, the Banwell Committee proposed engaging the main contractor and specialist firms early in the design and planning stages to anticipate difficulties and reduce the number and value of variations and claims. Banwell also called for all payments to be made "regularly and promptly," something that contractors and subcontractors have never quite managed to fully achieve in spite of legislation to that effect.

The organisation and management of training was one of the subjects dealt with by Banwell. Excluding the trade union side and the interests of the construction workforce is a recurring weakness of many of the reports, and Banwell was supportive of consolidating all operative training in the CITB. He recognised the need to train both operatives and managers. The report considered industrial relations, the role of shop stewards, temporary employment and worker registration, though not all of their discussions were fruitful. As Langford (2003, p. 103) was able to say, "the report heralded changes in the industrial relations climate and practice in industrial relations on large industrial sites. As a report it was farseeing and imbued trade unions and employers with a liberal approach to industrial relations." However, the Banwell Report was published in the decade immediately prior to the Thatcher era, which must have reduced its longer term impact.

The construction industry is a labour intensive industry and is a major employer. In 1970 the Large Industrial Sites Report was one of a minority of reports that engaged with the trade unions as well as the employers and professional bodies. Relations between the trade unions and employers had become so hostile by the late 1970s that involving the trade union movement in discussions about the future of the construction industry had become a political issue. This prevented an important part of the construction industry, namely the people who actually carried out the work on site, from participating in discussions on reform of the industry in several reports in this period.

The Wood Report

It was, however, not the final construction report before the industrial changes made under the Thatcher government of 1979 that extinguished the role of the unions in discussions on the way forward for construction, apart from the formal annual discussions on Working Rules Agreements, which kept up a pretence of negotiation between employers' organisations and the trade unions. In 1975 a report entitled "The Public Client and the Construction Industries" was published as the result of a collaboration between the social partners (namely government, trade unions and industry), under the chairmanship of Sir Kenneth Wood. The report is better known as the Wood Report.

It recommended, amongst other things, a minimum three year rolling programme of construction projects across the public sector to stabilise demand for construction regardless of other economic problems that confronted government and the economy. Wood argued that the search for the lowest priced tender was at the expense of the effective use of resources and improvements in efficiency and performance. The report was critical of the lack of time to prepare drawings and manage resources prior to commencing work on site. The result of these problems led to delays and variations, both of which in turn increased the cost of construction. The Wood Report was supportive of procurement methods such as lists of firms, restricted to firms known to be competent and have the capacity to carry out the work, two-stage tendering, and design and build contracts. However, like the Large Industrial Sites Report of 1970, the Wood Report did not concur with the approach of government in the 1980s and 1990s, which meant these proposals were not of lasting influence over government construction policy.

In 1977, according to Connaughton and Mbugua (2003), representatives from eight organisations came together to form the Group of Eight (G8). Members of the Construction G8 were the Institution of Civil Engineers, the Royal Institution of Chartered Surveyors, the Construction and Civil Engineering Group and Building Crafts Section of the Transport and General Workers Union, the National Council of Building Materials Producers, the Royal Institute of British Architects, the Federation of Civil Engineering Contractors, the National Federation of Building Trade Employers and the Union of Construction Allied Trades and Technicians. Although it was a single body that attempted to represent the diverse interests within the construction industry, it was criticised as ineffective, and in 1987 the G8 ceased to exist. Nevertheless, it demonstrated that it was possible for the different interests in the construction industry to come together in one body. Its major failing was that it lacked ministerial weight.

The National Economic Development Office

In 1979 the change of government occurred with the arrival of Mrs Thatcher as Prime Minister, a year after the publication of the National Economic Development Office (NEDO) report entitled "How Flexible Is Construction," which argued that there was a need for construction demand and capacity to be matched and in balance with the processes involved. In 1983 "Faster Building for Industry" and in 1988 "Faster Building for Commerce" were published. Largely concerned with industrial building, both reports sought to speed up the processes involved, including planning and construction. Although the report found contractors were capable of faster construction without necessarily increasing costs, it suggested that a "principal adviser" might be useful for less experienced clients, a role that is currently often taken up by project managers.

It suggested that, if construction were to be improved and speeded up in general, the quality of site management mattered, that communication between the designers and the clients, especially in a co-operative atmosphere, could avoid delays, and that a clearly defined programme of work with incentives, if helpful, could all be employed to speed up construction. If information was incomplete, then two-stage tendering or negotiation could be usefully employed. Design and build and management contracting were also recommended as useful procurement systems for speeding up the construction process. As a generalisation, the main culprits responsible for project delays, according to the NEDO report, were the statutory authorities, not the contractors, the clients or the architects.

In order to cut the duration of the construction phase, "How Flexible Is Construction" suggested the use of prefabrication and the standardisation of building components instead of relying on site labour. It also advocated the increased use of plant and equipment on site. At the same time the report criticised the poor image of the industry for deterring potential staff from seeking employment in construction, but it did note the pride taken in their work by the highest skilled workers and craftspeople. At the same time the report argued that higher education institutions should be more accessible to encourage more training and education in construction trades and management skills.

Male (2003) points out that the report "Building Faster for Commerce" identified a number of critical project success factors, including effective team working, clear briefing by clients and clear contract terms. At the same time as a building develops its design, buildability also evolves in terms of complexity confronting the management of the process with more issues to deal with.

The Latham Report

The recession in the construction industry at the end of the 1980s and early 1990s had led contractors to adopt a strategy of winning work by bidding low in the hope of winning and then making claims when variations were agreed, turning a loss-making bid into a profitable project. This meant construction projects were frequently places of argument and conflict between contractors and their subcontractors and between contractors and their employers. Hence when the Latham Report was published in 1994, Latham called for greater harmony and collaboration to improve the relations between the participants in the construction process and raise the performance of the industry.

Latham's report, entitled "Constructing the Team," was published in 1994. One of the main features of his report was the setting of a number of targets, including a 30 per cent productivity improvement. Unlike so many other reports on the construction industry, the Latham Report was followed up with an act of parliament, the Housing Grants, Construction and Regeneration Act of 1996, which attempted to solve issues that had been raised in the report, including payment delays and disputes. Latham also called for a revision of the standard construction contract together with the adoption of partnering agreements between firms to improve working practices. Latham favoured the New Engineering Contract (NEC) over the Joint Contracts

Tribunal (JCT) family of contracts, but both sets of contracts continue to be widely used more than 25 years after the Latham Report.

The report also recommended partnering as a means of improving relations between employers (or contractors' clients) and contractors and encouraging greater participation in the construction process by employers, while encouraging main contractors to work more closely with their supply chains. It remains questionable whether partnering has managed to overcome the distrust that pervades the building industry, and it remains to be seen if it has the means to do so. Though Cahill and Puybaraud (2003) assert that the use of partnering agreements continued to increase after the turn of the century, their continuing popularity may be more due to the spreading use of IT software, known as Building Information Modelling (BIM), which requires close collaboration between all parties involved in the construction process. Indeed, Cahill and Puybaraud (2003, p. 159) note that senior representatives of firms in the industry appeared very dismissive of Latham as late as 1999, claiming the Latham Report "had very little, if any, effect on the construction industry."

Nevertheless there have been a number of changes in the building industry, possibly brought about by changes in the technology used in construction and the use of IT in the form of BIM. These changes may well have influenced the way firms collaborate and interact in the management of the building process. These changes have also occurred as ever in response to changes in the wider economic climate and the business environment. Other changes in construction range from increasing the use of prefabricated components to the use of adjudicators instead of the courts to resolve disputes.

The Technology Foresight Report

In 1995 the Technology Foresight Report "Progress through Partnership" compared the construction industry and the offshore oil and gas industries. The comparison is based on a false premise, as Green (2003) points out. The two industries, he argues, are significantly different in character. The offshore industry is dominated by a few large companies and high barriers to entry, giving the established firms high profit margins. This is completely the opposite of the construction industry, which comprises a large number of small firms competing in a market with low barriers to entry, resulting in extremely low profit margins. With high profit margins in one and low profit margins in the other, it is not surprising to find investment in research and development is much greater in the offshore oil and gas industry than in construction. Although the Technology Foresight Report called for improvements in "productivity, cost and quality," Green points out that there were no suggestions on how these improvements might be achieved.

The Egan Report

In all of these reports on the construction industry up to this point, there was little or no attempt to analyse the cause and effect relationships between the business environment and the construction industry. This changed with the Egan Report of 1998, when several drivers of change, though not necessarily the causes of change, were mentioned. Murray (2003) lists Egan's drivers of change in the construction industry. First is committed leadership, followed by a focus on the customer. The third factor is the need to re-engineer the process by integrating the process and the construction team around the product. The next factor is the need for an agenda for change built on quality by delivering projects punctually, without defects and on budget. Lastly, Murray identifies Egan's driver of change as a commitment to people, involving not only their working terms and conditions but also their training and development or empowerment. The Egan Report advocated that qualifications should be included on identity cards to be presented to employers. To some extent progress has indeed been achieved in this area since the Egan Report, with the introduction of the Construction Skills Certification Scheme (CSCS) card. However, how these drivers of change actually alter the way the industry operates remains mostly vague and hidden.

Unfortunately, this attempt to understand the industry is weakened by the setting of arbitrary targets for the industry, including at various times:

- a 10 per cent reduction in capital costs
- a 10 per cent reduction in construction time
- a 20 per cent increase in predictability (based on budget and delivery)
- a 20 per cent reduction in the number of defects
- a 20 per cent reduction in the number of reported accidents
- a 10 per cent increase in productivity (measured in terms of value added per head)
- a 10 per cent increase in turnover and profits of construction firms

There is no attempt in any of the reports that use targets to show how these target figures were arrived at. Nevertheless, following Egan, several reports continued to set arbitrary targets without any transparency or rationale. Scientific method and intellectual rigour were also found wanting when the task force under Egan asserted that there was no difference between construction and manufacturing and, like manufacturing, construction could be broken down into four areas, which were product development, project implementation, partnering suppliers and producing components. Unfortunately, the manner in which those elements of production are achieved in construction is very different from general manufacturing, such as automotive car production.

For example, project implementation in construction is a fragmented process, with the main contractor taking the lead to co-ordinate several specialist independent subcontractors. While firms in the construction process co-ordinate their activities, they rarely work closely together due to lack of trust. The production of components is carried out by separate suppliers, whose contribution is defined by contractual arrangements rather than a collaborative ethos. As a result the whole construction process is highly fragmented.

In construction projects, product development is largely concerned with design and takes place independently of the contractors. Just as Banwell had criticised the separation of design from construction in 1964, Egan now advocated the early involvement of contractors and subcontractors to integrate the design process with the construction processes to follow. This was to be extended to demonstration projects that were to be designed to disseminate construction innovations. Murray (2003), however, points out that many projects show no clear innovation apart from those introduced by suppliers. Taking a broad definition of construction, it can be argued that construction product manufacturers are often where research, development and innovation are located in the construction industry, not amongst contractors and subcontractors, unless they are supply and fix contractors working on behalf of a product manufacturer.

Egan advocated the use of key performance indicators (KPIs). Of the KPIs used in construction, seven relate directly to projects. They are construction cost, construction time, cost predictability, time predictability, defects, product satisfaction and service satisfaction. Three KPIs measure aspects of company performance. They are profitability, productivity and safety. Unfortunately, construction firms tend to avoid publicity and the risk of criticism in case it is used to dent their own reputation, even when the data is anonymised. Under these circumstances it is difficult to take KPIs as a serious measure of industry performance. Perhaps this is why even Egan (2001) admitted, "Normally what happens to Government reports is people flick through them and throw them in the bin."

The Fairclough Report

Nevertheless, following the Egan Report, a Minister of Construction and sometimes also a Minister of Housing were appointed with frequent reshuffles. In spite of the size of government expenditure on projects involving construction, the role of these Ministers became one of a portfolio of responsibilities of the minister concerned.

Egan did recognise the role of operatives as members of the construction team, a point reinforced by Fairclough (2002), whose review on research and innovation in construction made several recommendations, including maintaining investment in construction R&D that might yield productivity improvements, while developing skills in the industry. The Fairclough Report also recommended that research in construction could be used to encourage the industry somewhat idealistically to develop a broad vision for its contribution to what it called the quality of life agenda.

Picking up on the Fairclough Report, the Department of Trade and Industry (DTI) published a report in 2007 entitled "The Construction Research Programme – Project Showcase." This document reported on government sponsored construction research projects and revealed the priorities of public sector research funding policies. These priorities included economic benefits such as supporting a profitable and competitive industry both in the UK and abroad. A second set of priorities was to provide customer satisfaction and value, meeting the needs and expectations of all stakeholders, improving profit margins and performance, ensuring respect for people and helping them to develop their skills and careers. The social benefits of construction research should also be concerned with providing a safe and healthy built environment. A third set of priorities was concerned with creating environmental benefits with a construction industry that took the environment into account and reduced the consumption of natural resources and the use of energy.

The Wolstenholme Report

Of all the construction reports since the Second World War, the Wolstenholme Report (2009), "Never Waste a Good Crisis," is perhaps the most systematic attempt to explain the behaviour of firms in the construction sector in order to derive policy proposals. Reflecting on the Egan Report, Wolstenholme (2009) wrote that there was disappointment at the slow progress made in adopting Egan's proposals. Wolstenholme went on to provide an understanding of the slow rate of change in construction based on a survey from the point of view of the "diverse constituencies of UK construction." He reported that there was a lack of resolve to implement the changes advocated by Egan, identifying what Wolstenholme described as "blockers" or impediments to change. There were four blockers in all. They were business and economic models, capability in the form of high calibre staff at all levels, the construction delivery model and, finally, the structure of the industry itself.

One reason for the first blocker, namely the slow rate of change in construction, according to Wolstenholme, is because of the lack of incentives for contractors. The shortage of high calibre staff, the second blocker, was because of the poor image of the industry and the lack of inspiring public figures to represent the industry in public. The third blocker was due to the fragmented structure of the construction industry. The many trades and professions prevented a more accurate picture of the true size of construction, its importance and significance from being understood. Few, if any, viewed the industry as a whole from beginning to end, from architects and materials suppliers through specialist suppliers and main contractors, to facilities managers and property management. Finally the fourth blocker referred to the lack of vertical integration within firms in the construction industry, which limits the ability of the industry to innovate and modernise.

Under the heading "Big Themes for Future Action," Wolstenholme suggested a number of challenges or propositions, including the need to find a coherent representation of the construction sector, calling for industry bodies and professional associations to collaborate to enable the industry to represent itself to government and others. He also argued that there is a choice to be made: either the industry continues to use lowest-price tendering and litigious relations or it seeks to enter into collaborative methods of working to improve the industry, its output and services. In fact, Wolstenholme makes the very valid point that when Egan compared construction to the automobile industry, he did not mean to imply that construction was like car manufacturing but that there were lessons that could be usefully learned by contractors from the best practices in manufacturing industries.

The Farmer Report

One of the most recent reports on the construction industry, the Farmer Report (2016) entitled "Modernise or Die, The Farmer Review of the UK Construction Labour Model," consciously adopts a medical analogy in analysing the problems of the industry. Firstly, the report identifies the symptoms, then diagnoses the causes and goes on to make a prognosis. The next step is designing a treatment plan for recovery, and the report ends by suggesting the industry should be kept under observation. In the foreword, the report draws attention to "the conclusion that – given workforce attrition exacerbated by an ageing workforce – we simply cannot go on as we are." Using a health metaphor implies that the industry is in ill health and requires treatment to bring construction back to being fit for purpose. In the introduction to the report it promises to find the causes of the malaise of the construction sector.

The report identifies "[the] critical symptoms of failure and poor performance," repeating the issues confronting the construction industry that were identified in previous reports, going back to the Simon Report of 1944. The list of symptoms includes low productivity and unpredictability, fragmentation of the structure and leadership of the industry, low profit margins and a lack of a co-operative and improvement culture, and finally the poor image of the industry is noted.

The Farmer Report at best reiterates that little progress has been made to deal with these problems over the decades. The report attempts to argue that there are three root causes. Firstly, Farmer claims that in response to the business environment of low capital reserves and volatile demand, construction firms have adopted a strategy to enable firms to merely survive. Secondly, according to Farmer, traditional procurement methods persist, and thirdly, there is no strategy in place to modernise the industry. This state of affairs is now accepted as the norm. There is a "lack of government policy" to deal with market issues in construction, including the volatility of demand, resistance to change and closer working together with clients.

He recognises that the industry does not have the means to reform itself. Reform of the construction industry will have to come from the demands and requirements of clients, or through government intervention, strategy and regulation. Farmer (2016, p. 10) states, "At the heart of this review's recommendations . . . it is proposed that a new . . . tripartite covenant is established between the construction industry, its end clients (private and public) and government acting as a strategic initiator."

The Farmer Report makes several recommendations; including reforming the Construction Industry Training Board (CITB), but it is not clear what should emerge from the changes. Industry, clients and government should also collaborate to increase investment in R&D and innovation and shift production towards prefabrication. Although it recommends collaboration in innovation, the report is unclear about what and how that might be achieved. Government industrial strategy, according to Farmer, needs to recognise the value of the construction sector and the education and training needed to reinforce its strategy. Government should develop a comprehensive pipeline of housing demand similar to the National Infrastructure Pipeline.

Farmer attributes poor productivity in construction to a fragmented and complex organisational structure between clients and contractors, poor client briefing and late changes to the brief by clients and the frequent need to re-work and rectify defects. However, the report does not say who is responsible for low productivity, who is to blame and how it can be remedied. Under these circumstances many of the causes of low productivity are beyond the control of the contractors, who are only attempting to meet the demands and requirements of their clients.

A key point highlighted by Farmer (and referred to, quite independently, in the Egan Report of 1998 and the Wolstenholme Report published in 2009) is the problem of low profitability of construction, which he rightly says has been a long standing problem in construction. In fact more accurately it is the operating margins of contractors that are an underlying cause of many of the problems in construction, together with low productivity. Farmer (2016, p. 23) notes that this key accounting measure of performance, namely operating margins, had declined from "2.8% in 2010 to an average of 1.2% in 2013." This refers to the average net annual profit margin of the top 100 contractors, namely the percentage profit they make on each project on average. Given the precarious nature of construction it is not surprising that the losses on one project can wipe out the profits of the firm generated from all of its other projects, making firms extremely risk averse when it comes to investing in new plant or initiatives but not when it comes to tendering for work. The low profit margin in the construction industry is the single main underlying long existing cause of the poor performance of the sector.

Construction strategy – Construction Sector Deal

In the most recent publication in this review of reports the government itself recognised the unsatisfactory state of the construction industry. According to the Construction Sector Deal (2018, p. 32), "The current business model of the construction sector is not sustainable. Construction customers and businesses across the supply chain are focused on the costs and risks of individual projects, and do not collaborate effectively."

The report also highlighted poor productivity in the construction sector and rightly attributed this partly to the low levels of investment in research and development by contractors, amounting to £211m in 2016 compared to £3.3bn by the automotive industry and £1.9bn by the aerospace industry. However, the turnover needed to generate £211m of R&D investment by contractors is a great deal more than the sales needed by an automotive or aerospace company to meet the same investment requirement. These comments regarding the poor construction business environment and the R&D challenges facing contractors implied that the government was determined to improve the situation facing firms in the building industry.

To that end the Construction Sector Deal reported several measures that were being taken by government in an effort to strengthen the building industry. For example, the report mentions the National Infrastructure Delivery Plan 2016–2021 and the infrastructure pipeline, which was planning a ten year programme of investment in infrastructure with a plan or pipeline of work expected to be worth £600bn. This scale of planning is to be welcomed as it gives contractors the opportunity to plan and hire staff with a degree of certainty.

Another welcome move by government is to reinforce the legislation regarding payments by main contractors whose turnovers are greater than

£36m per annum, or whose balance sheets are greater than £18m or who employ more than 250 people. In the future, according to the Sector Deal, such firms will have a formal Duty to Report on Payment Practices and Performance. It remains to be seen how well this legislation will work in practice, but it may well be of use to subcontractors, who frequently incur major problems waiting for payment.

Other aspects of the industry covered by the Sector Deal include health and safety, apprenticeships and training, digitisation and off-site production and whole life asset performance. It is, however, not clear how they may all be combined in an integrated strategy for the industry. Unfortunately, the Sector Deal adopts the *Construction 2025* report (2013) to repeat the call for a 33 per cent reduction in costs, a 50 per cent reduction in time to build, a 50 per cent reduction in greenhouse gas emissions and a 50 per cent reduction in the trade gap, all old targets that have never been met and are discredited by many in the industry for reasons covered in Chapter 3.

Concluding remarks

This overview of reports published in the last 70-odd years has highlighted a number of key points. Very few reports have sought to look at the economics of the industry to try and explain the reasons for, and causes of, the way the firms in the industry behave. Without looking at the causes it is difficult to see how any of the reports could come up with viable solutions. The approach adopted in these reports largely relies on intuition and experience. The method of analysis is usually very weak. Nevertheless, taken as a whole, the reports are peppered with practical solutions, and much can be gleaned from them. They tend to offer a wish list of measures without any clear indication as to how they may be achieved.

The time has come for a fresh approach to dealing with the problems of the construction sector. In an industry composed of so many small firms but dominated by only a relatively few large players, government intervention in some areas of construction, most notably in training, planning and public sector demand management, is essential if the industry is to improve the quality of the services it provides and go on to prosper in its own right.

The economics of the construction industry should inform the direction of change the building industry might adopt to resolve the many issues it faces, including skills shortages, disputes, productivity in the industry and many more challenges. The proposals here emerge from an economic understanding and perspective of the construction industry.

It is the realm of economics that provides the insights into the production processes, conflicts that arise, relationships between the players, their interests and the problems confronting the contractors and their employers.

In the rest of this book an attempt is made to understand and explain the economics of the construction industry in the hope that workable solutions can be found and suggested that might conceivably be adopted. It would be a rather nice legacy to leave behind from a career observing the industry from the outside in academia.

Bibliography

Banwell, Sir H., (1964) *The Placing and Management of Contracts for Building and Civil Engineering Work*, London, HMSO

Cahill, D. and Puybaraud, M.C., (2003) "Constructing the Team: The Latham Report (1994)," in *Construction Reports 1944–1998*, eds M. Murray and D. Langford, Oxford, Blackwell

Connaughton, J. and Mbugua, L.A., (2003) "Faster Building for Industry: NEDO (1983)," in *Construction Reports 1944–1998*, eds M. Murray and D. Langford, Oxford, Blackwell

Department for Business, Energy and Industrial Strategy (2018) *Construction Strategy: Construction Sector Deal, Department for Business, Energy and Industrial Strategy*, London, BEIS, Open Government Licence

Designing Buildings Wiki, (2017) *Construction Industry Reports*, www.designing buildings.co.uk/wiki/Construction_industry_reports

Egan, J., (1998) *Rethinking Construction: Report of the Construction Task Force*, London, DETR

Egan, J., (2001) "Egan on Egan," *Building*, 10 August, pp. 18–19

Emmerson, Sir H., (1962) *Survey of Problems Before the Construction Industries: Ministry of Works*, London, HMSO

Fairclough, Sir J., (2002) *Rethinking Construction Innovation and Research*, London, Department for Transport, Local Government and the Regions

Farmer, M., (2016) *The Farmer Review of the UK Construction Labour Model: Modernise or Die, Time to Decide the Industry's Future*, London, The Construction Leadership Council

King, V., (1996) "Construction the Team: A.U.S. Perspective," in *CIB Working Commission 65 Organisation and Management of Construction Symposium, Conseil International du Bâtiment*, ed D. Langford, London, E & FN Spon

Langford, D., (2003) "Large Industrial Sites Report (1970)," in *Construction Reports 1944–1998*, eds M. Murray and D. Langford, Oxford, Blackwell

Latham, M., (1994) *Constructing the Team – Final Report of the Government/ Industry Review of Procurement and Contractual Arrangements in the UK Construction Industry*, The Department of the Environment, London, HMSO

London Cabinet Office, (2011) *Government Construction Strategy*, London, The Cabinet Office

Male, S., (2003) "Faster Building for Commerce: NEDO (1988)," in *Construction Reports 1944–1998*, eds M. Murray and D. Langford, Oxford, Blackwell

Moodley, K. and Preece, C., (2003) "Survey of Problems Before the Construction Industry: A Report Prepared by Sir Harold Emmerson (1962)," in *Construction Reports 1944–1998*, eds M. Murray and D. Langford, Oxford, Blackwell

Murray, M., (2003) "Rethinking Construction: The Egan Report (1998)," in *Construction Reports 1944–1998*, eds M. Murray and D. Langford, Oxford, Blackwell

National Economic Development Office, (NEDO), (1983a) *How Flexible is Construction*, London, HMSO

National Economic Development Office, (NEDO), (1983b) *Faster Building for Industry*, London, HMSO

National Economic Development Office, (NEDO), (1988) *Faster Building for Commerce*, London, HMSO

Simon, S.E., (1944) *The Placing and Management of Building Contracts: Report of the Central Council for Works and Buildings*, London, HM Stationery Office

Wolstenholme, A., (2009) *Report: Never Waste a Good Crisis: A Review of Progress Since Rethinking Construction and Thoughts for Our Future*, London, Constructing Excellence

Wood, Sir K., (1975) *The Public Client and the Construction Industries: The Report of the Joint Working Party Studying Public Sector Purchasing, Building, and Civil Engineering Economic Development Committees*, London, HMSO

2 A critique of the modern construction industry

Although the Construction Sector Deal of 2018 included setting industry-wide targets, including cost reductions, time reductions and greenhouse gas reductions, setting arbitrary targets for the construction industry has become outdated and irrelevant. They have proved to be ineffective in achieving their objectives, as there is no penalty for not achieving them, and there is no organisation to take responsibility. Moreover, the non-attainment of the targets only serves to throw another criticism at the construction industry. The time has come for an approach to strategic planning for the construction industry that replaces targets with priorities.

In 2011 the Cabinet Office, based in Downing Street, published a report on the construction industry criticising the failure of the government itself to drive growth and "exploit the potential" gain from construction procurement. The report, entitled "Government Construction Strategy," foresaw somewhat wishfully that the adversarial culture in the construction industry would be replaced with a collaborative philosophy. However, the report did not mention how this would be achieved. It is possible that the authors of the report had in mind the introduction of Building Information Modelling (BIM), which was emerging at the time as a new state-of-the-art application of information technology to construction project management. BIM, it was hoped, would enable firms to work collaboratively on projects throughout all phases of each project's life cycle from design, through construction, to the running and maintenance of the project.

However, the 2011 Government Construction Strategy omitted to mention how the difficult problems of sharing information between firms working on a project might be overcome. BIM requires a degree of information sharing that firms were not always happy with as they considered the information to belong to them, their trade secrets, which gave them authority and bargaining power within the network of firms that made up the supply chain. The report avoided discussing the lack of trust between firms in the construction industry, the effect of conflicting interests between the separate firms in the supply chain and the use of power by main contractors in the relationship between main contractors and their suppliers, a power used to put pressure on subcontractors to deliver at ever lower profit margins, or to increase the credit period subcontractors were forced to wait for settlement of their accounts.

Innovation and construction

The Government Construction Strategy of 2011 emphasised the need for innovation in the supply chain. The supply chain, from a main contractor's point of view, comprises all those firms engaged on a project by specialising in the different trades hired to build, including specialist firms with specific skills, plant hire firms and those firms that provide building components, materials and products. However, while calling for innovation the report did not appear to appreciate that the process of innovation in construction occurs on every project every day. Innovation in construction takes place every time a new product, material or process is introduced on site. Far from being unengaged with new processes, materials or techniques, construction firms are continually innovating on site in a practical way. This occurs so frequently that it is taken for granted. Nevertheless, while very difficult to measure, these constant improvements have the effect of increasing real productivity, though not always captured or measured in monetary terms.

If innovation is the adoption and spread of new products and processes previously tried by other firms or in different circumstances, then invention is the creation of something entirely new that has not yet been attempted. It is the first time a new product or process is tried. Further copies of the original idea, especially by others, are gradually then adopted throughout the industry, raising productivity and efficiency. One of the hidden benefits of the fragmented construction industry, in which several small firms combine to produce the built output, is that firms observe what other firms are doing and, when they see a new technique being used successfully, will attempt to adapt it to their needs. For example, the invention of drones is a combination of existing technologies, which are used to oversee building sites without the need to necessarily visit the particular location on site in person.

Truly inventive new products and materials that represent a technological breakthrough are rare, and even when they do occur, it can take several years before they are incorporated into new products or services. For example, graphene represents a step change in materials science at a molecular level and is a modern invention rather than a new application of existing materials. Graphene was first found in the 1960s and took until the beginning of this century before its potential was recognised. Similarly, the history of the laser goes back to a paper written by Einstein in 1917. Only in the 1980s did developments in laser technology open up new possibilities that were barely imagined at the time, applications ranging from holograms to facilitating several medical procedures. In the building process, lasers are used to monitor inventories and identify particular components, not to mention several technical applications.

Although significant inventions are relatively rare, innovation is common and occurs on a daily basis in construction as new processes are constantly being introduced on site. If innovation is the spread and adoption of new ideas or methods of working, firms in the construction industry are highly innovative out of necessity.

To understand the impact and significance of innovation in construction one needs to appreciate that modern construction is a complex process, using a multitude of diverse processes and many products and technologies. As a result any one of these products forms only a small fraction of the total cost of building. Even when a new product, material or method results in a large percentage saving of its own cost, it will have only a minimal impact on the overall cost of a building, because in itself it is only a small proportion of he total cost of the building. When every new product is introduced, it must be integrated into the building process, fitting in with established products and existing materials.

The success of the innovation also depends on the abilities of the skilled and semi-skilled workers on site. Firms continually adapt their building techniques in order to accommodate new ideas as and when they appear on site or on the drawing board, each time improving productivity. Building firms are invariably on a steep learning curve, adopting new ideas and adapting existing components and materials to combine new ideas and methods with older techniques and technology. That is the nature of the building industry as far as individual firms are concerned.

Construction contractors have few research and development departments as such, except in some of the largest firms. Even a leading firm, such as Laing O'Rourke (2017), uses its R&D department in the form of its Design for Manufacture and Assembly (DfMA) facilities to collaborate with consultancies and universities. University departments are often at the forefront of research and development in construction. One major research facility is the Building Research Establishment (BRE). It also carries out research, which it uses to promote and spread ideas in construction. Originally a public sector research facility, it has since become an independent research establishment concerned with building performance, standards and the environmental impact of built assets.

However, to identify where invention and innovation are taking place in construction, one needs to look beyond most construction firms. Much invention and innovation in construction actually take place, though not exclusively, in the R&D departments of construction product manufacturers.

It is they who compete with other product manufacturers by bringing out improvements to their product lines. These improvements may appear to have little financial or economic impact, but each one is an improvement, because it increases productivity, saves time in the construction programme, saves labour on site or uses less energy. Taken as a whole, changes in construction techniques form a continuous process of improvement in construction. Those who accuse the industry of not being innovative are simply looking in the wrong place. Every project offers a challenge in terms of innovation. Every project is its own research and development platform, on which firms have little choice but to innovate.

Far from being backward, innovation in construction is an ongoing process. There is a need to recognise that research and development of new products and techniques take place not necessarily on construction sites but in the labs and on the shop floors of the manufacturers of construction products and materials. New products and materials are being introduced on site on an almost daily basis. However, their impact on the construction process can be appreciated only in the long term as any one change or innovation tends to be insignificant in terms of the overall project and affects only specific trades and specialisms. Rarely are the changes noticeable in the short run, but over time, taken as a whole, the changes have transformed and continue to transform the way construction is carried out. Today, far fewer people than in the past work on site as more and more of the construction process is prefabricated and mechanised, and off-site manufacturing and assembly are simplified and speeded up. These changes have come about because they have economic benefits for the firms introducing them, while satisfying the needs of the supply chain and their customers. Technology drives change in the construction industry just as it does in all industries.

Legislation and economic conditions also bring about change in the way the construction industry operates. For example, competition from abroad forces firms to find ways of reducing their prices, while legislation forces firms to comply with legal requirements. For example, the Construction (Design and Management) Regulations 2007 (CDM) introduced codes of behaviour on health and safety that have altered the attitudes, behaviour and training of the workforce on site, at least on larger projects and probably also on the vast majority of sites across the construction industry.

The low profit margins in construction contracting

Innovation is used by firms to reduce their costs, and this theme was taken up in the Government Construction Strategy Implementation Report of 2012, which stated that the overarching aim of government was to reduce construction costs by 15–20 per cent. Adopting such a high savings target for construction firms almost implied that contractors were overcharging their clients. Far from overcharging their clients the opposite is nearer the case. The annual reports and accounts of construction contractors show that profit margins of construction contractors are extremely low relative to profit margins found in other sectors of the economy. In construction profit margins in the annual accounts of firms are often around 2–3 per cent, sometimes even less.

These very low profit margins imply that construction firms survive by being extremely competitive in their bidding for contract work. Each competing contractor is aware that its competitors are also pricing for the work, using similar materials and suppliers all charging similar amounts. Each firm prices as low as it can, given the risks of under-pricing, especially if costs rise unexpectedly during a project. When the bid is so low that the contractor incurs a loss, this is referred to in game theory as the winner's curse.

Because their profit margins are so small, construction firms have to undertake large amounts of work so that the profit they make generates a sufficient return in total even though the rate of return is low. In other words, relative to their capital assets, they have to win a large amount of work in terms of project value in order to generate enough profit from each project to build up an adequate return on capital, a process referred to as sweating the firm's assets. This high risk business model enables firms to achieve a total mass of profit that is just an acceptable return on capital for their shareholders. This amounts to a very risky strategy for building firms. Consequently, firms in the sector are vulnerable and frequently go to the wall.

Tight profit margins are imposed on contractors by the competitive tendering process, which assumes each firm is offering an identical project and quality of service as that of their competitors. This leaves contractors little room for manoeuvre apart from price. In order to win work firms attempt to undercut their competitors by cutting their own profit margins to the bone. A cost overrun or a delayed payment can be fatal for a contractor or subcontractor, and therefore each and every project a firm undertakes in construction is simultaneously an opportunity to make a contribution to profits as well as a threat to the continued existence of the firm. This is the case for the largest firms as well as the vast number of small and medium sized enterprises that make up the majority of firms in the construction industry.

Firms in construction frequently survive due to the skills of their managers in handling cash flow, often at the expense of their subcontractors. The harsh reality of cash flow management with frequent delays in payments to subcontractors results in conflicts, disputes and poor relationships between main contractors and their subcontractors and suppliers. The very low profit margins found in the construction industry mean there is very little room for error, negotiation or concession for the sake of goodwill, especially when disputes arise. Low profit margins force firms to expand turnover to such an extent that the level of risk and default is greater in construction than in almost any other sector of the economy. The high risk faced by contractors working with low margins can easily result in firms' profit margins being eroded altogether on individual projects. As a consequence the construction industry's reputation is impacted by its highly litigious and confrontational nature brought on by its own level of conflict between the different building contractors.

For many firms in the construction industry the main motivation is short term survival in a very competitive environment. It is the market that shapes what they do and how they behave. Contractors are aware of the intensity of competition through the number of competitors they face and their own losing bids for work in the tendering process. Contractors can respond to demand only by tendering for work to build up their turnover and cash flow. Building contractors themselves cannot create demand. Demand for construction in general is determined by the subjectively perceived need for buildings, which is based on clients' decisions to build. Construction demand is not determined by the cost of construction. The pricing of construction work is important but is primarily used to select the winning contractor in the tendering process. The final cost of construction is derived from the actual work carried out, which, given the degree of uncertainty at the start due to incomplete information, unfinished designs, delays during construction, lack of decision making by the client and variations in design during the construction phase, is often greater than the cost anticipated before work on site has commenced. The low cost of expected building work is caused by the optimistic bias in costing projects in order to satisfy the public committees and boards of company directors, whose members must provide approval for a project to go ahead. This is often seen in public sector projects, which understate the expected cost of the projects. There are innumerable examples of infrastructure projects and public sector buildings where the initial cost estimates have been overtaken by the outturn costs due to this optimism bias. Often the buildings would not have gone ahead if the actual high outturn cost had been used for decision making purposes. This does not mean that even at the higher outturn cost the buildings should not have been built or the money was misspent. Indeed buildings cost a large lump sum, but spread over the life of a structure these costs are negligible compared to the labour, running and maintenance costs of a building over its entire life time.

We have therefore seen that construction cost uncertainty comes from all the unanticipated problems and difficulties that can arise in the course of a project, ranging from adverse weather to unexpected ground conditions, design delays, accidents, late deliveries and disputes. In the private sector similar uncertainties may also come from a bias in the pricing of projects by developers in order to have decisions to invest agreed by funders. They also come from contractors themselves under-pricing the work, in order to win the tendering process. In this complex market of construction project delivery the government target setting approach may be appealing in its simplicity, but targets do not in any way offer a solution to the problems that emerge in construction.

For example, a government target of construction cost reduction does not take into account that the role and function of price in construction differ in several ways from many other industries. In manufacturing, for example, reductions in cost tend to be passed on to consumers over time through competition. Consumers respond to lower prices by increasing the overall quantity they purchase as a whole group of buyers. For example, this can be seen most clearly in the market for personal computers and laptops. As the price of computers declined, sales expanded to form a mass market for the producers, out of which the lower profit per unit sold was more than compensated for by the large number of sales experienced by manufacturers. In contrast to other industries, in the built environment sector lower construction costs and prices do not lead to increased demand or higher numbers of units sold over the longer term. All that happens is that land prices increase.

The implications of lower building costs

The influence of government on the construction sector is partly because the public sector is a major component of construction demand. As a client, it influences how the construction sector behaves, just as a good teacher can influence the behaviour of a class of pupils. Firms have to comply with government wishes if they wish to be considered for work in the future. The government has the potential to determine the manner in which public sector procurement is to be conducted. If the government chooses not to do so, it cannot blame the construction industry for being unruly and not changing its ways any more than a teacher can blame his or her class for being out of control. A government procurement strategy is essential if the government wishes to influence change. Firms tend to rise to the challenge if they see market opportunities.

Unfortunately, the government's strategy has focused on reducing building costs without showing an understanding of the relationship between the cost of construction and the price of a finished building. Indeed the two are barely connected. In general construction costs are not the single major cost of building: land is. The cost of a building is the combination of a site plus construction work, but the cost of construction is not taken into account when purchasing a finished building, nor is the cost of land. The value of a building is worth only what a purchaser or tenant is prepared to pay, usually a price equivalent to similar buildings in the area. That said, the value of a house depends on the number of people interested in buying it, their confidence that house prices will not fall and their purchasing power or income. As in any auction, the person or company willing to pay more than anyone else wins the bidding and the building. These concepts are well understood by the development community, who use terms such as residual land value and whole life costing to evaluate proposals and not just the capital expenditure of constructing a building, which nevertheless seems to be the focus of so many construction strategies up to the present.

One commonly accepted method used to value building proposals is the discounted value of the stream of future rents, real or imputed, and the rate of return the building purchaser is willing to accept. If construction costs decline, land market values rise as developers compete to buy sites. A consequence of this competition for sites is that lower construction costs are not passed on in the form of cheaper buildings but instead are used to increase land values as developers can dig deeper into their own pockets thanks to cheaper construction methods. The actual value of the building depends on rental streams after completion, not the cost of construction.

The target of cost reduction in construction strategy reports and the pressure on firms to reduce their costs result in a downward pressure on the wages paid to the construction industry workforce and reductions in the quality of working conditions and training. It is far less risky for contractors to reduce wages than to invest in new plant and machinery. It is much easier for firms to reduce the wages they offer to their workers than to reduce the prices paid for materials or other costs, such as interest paid to banks, over which they have no control.

Ultimately the performance of the industry depends on the quality and performance of the workforce and how well it is led by its managers. There is a need for fully funded training to raise the standard of building qualifications, which would require entrants to the industry to complete a period of training that was genuinely geared to high standards of competence. Indeed the ultimate objective of improved training and education should be the creation of skilled building professionals. This would lead to an enhanced status based on actual build quality and in turn give the UK construction industry an international appeal that would enhance the reputation of UK-based construction firms.

Instead, training needs are often neglected. Consequently, construction labour is frequently unqualified or under-qualified for many of the tasks it is set. Labour is often casually employed on a project by project basis, with very few working for the full construction duration of a project or continuing to work for the same employer once projects come to an end. There is little loyalty towards workers and little commitment in return. The small profit margins of contractors mean firms cannot risk taking on people who are not engaged in value adding work each and every day, and firms cannot afford to employ people directly or train them, especially if there is no guarantee of work or of retaining trained workers if they can command higher wages elsewhere.

The extreme lumpiness of the workload of construction companies is caused by projects that require a large and skilled workforce for discrete periods, for the duration of the project, after which their services are no longer required. This feature of construction calls for special measures if the problems that result are ever to be managed. The discrete nature of construction projects means that government has a duty to intervene in the market to mitigate unavoidable periods of unemployment or underemployment of the workforce.

The decoupling of construction demand and the price of buildings

The public sector operates in a completely different way compared to the private sector. While the private sector is concerned with business risk, returns and speculation, the public sector is driven by perceived needs, government policy and budgeted expenditure, often with political implications. The following discussion is concerned with the private sector only.

The clients of the construction industry hire contractors, because they need new buildings or structures or some repair and maintenance on existing buildings, not because structures have become cheaper to build. If one contractor wins additional work, it can only be at the expense of rival contractors who lose that particular opportunity to build. Competition between contractors is a zero sum game, meaning that one firm gains at the expense of another. The total size of the construction market does not change as a result of anything contractors may do. The size of the market (as measured by the total value of projects put out to tender or offered to contractors) is wholly determined by demand from developers and is influenced by factors such as interest rates, land prices, government policy and economic and business outlook, confidence and expectations.

The cost of construction may appear to be relatively high. Construction projects are indeed amongst the largest lump sum single purchases made by consumers, government or industry. However, because of the durability of buildings and the running costs of maintenance and staffing, the capital cost of building work is not a major barrier to investment for developers and governments, because these costs are spread over the life of a building.

Indeed, the initial cost of constructing a building is a relatively insignificant cost when spread over the structure's life of, say, 30 years or more, compared to annual staffing costs, for example.

Bearing these arguments in mind, reducing prices in construction, as the government advocates, only translates into lower turnover for construction firms, because it does not generate additional business or turnover. Economists describe the aggregate demand for construction as price inelastic – the quantity demanded does not respond sharply to changes in the price of construction. In other words demand for construction is not price sensitive. Hospitals, roads and schools are commissioned on the basis of government policy, departmental aims and budgets rather than in response to changes in construction prices. Similarly, private sector firms and developers invest in commercial buildings, provided they can sell or rent their completed projects. Investment decisions are taken on the basis of commercial advantage and market potential, which in turn are based on speculative asset values and financial returns, which are themselves based on expectations about the future. The cost of construction is only one element in a number of factors to be considered.

Apart from the cost of construction, land markets play a very important role in the construction investment process. This requires a brief explanation. The total cost of a building includes not only the cost of construction but also the cost of land. Property developers use a "residual method" for calculating the value of a site, the price of which they need to establish in order to purchase the land before building can commence. Table 2.1 demonstrates the residual method by showing what remains after the cost of construction and the developer's profit have been deducted from the gross development value (GDV) of the completed building. The GDV is the total value of the building, taking into account the annual future rents the building will earn less the cost of maintenance and other running costs. Firstly, developers estimate the GDV. In this example, let us assume the GDV is £10m, which is based on the present value of the combined total of all future annual rental receipts, either actually paid to the landlords of the building or implied, if the building is owned and occupied by the same firm or persons and no rents are actually paid. The developer then estimates the budgeted

Table 2.1 The residual method

Capital value of completed building based on the annual rental value (GDV)	£10,000,000
Cost of construction	£4,000,000
Developer's profit (20%)	£2,000,000
Residual	£4,000,000

cost of construction; in this case let us assume the cost of construction is £4m. Finally, the developer's own profits are calculated as a percentage of the building cost, usually around 20 per cent of the cost of construction, or £2m in this example. To calculate the residual value, the developer subtracts the cost of construction together with its own profit from the estimated capital value of the finished building, namely £10m less £4m less £2m. This difference between the GDV value of the building and the cost of construction and developer's profit is the residual value of £4m, which is then available to purchase the site.

Using this method of calculating the value of a site means the lower the cost of construction the more is available for the developer to set aside to bid for the site in the land auction. Any reduction in construction costs therefore benefits the land owner selling the land, not the developer or the construction firm. The financial value of a building is based on the rental value of the finished building, not its cost of construction. Unlike in other industries, contractors cannot increase their sales because the price of their finished goods, namely the buildings, is not made cheaper by lower construction costs, because any reduction in construction costs is counter balanced by the higher cost of land.

When the cost of construction is reduced through the introduction of new methods or materials, say, through innovations and improved technology, the savings made due to the lower cost of construction enable developers to increase their bids in the auction for land for building sites. Developers know that if construction costs have been reduced, competing developers will also have access to the same new technology, and each developer needs to sacrifice the gain from cheaper construction costs in order to reduce the risk of losing a potential site to rival bidders. This is a very gradual process, but over time the result is higher land prices, not cheaper buildings.

In practice, cost reductions favour those developers who already own land and are in a position to gain from reduced construction costs in terms of their own short run operating profits. In the longer term even those developers who gain from higher land prices will need to pay increased prices for land purchases due to cheaper building budgets. Otherwise, they would run the risk of losing the bidding war for a new site to a rival developer.

Naturally, the sequence of events may differ from this description from project to project. Developers often purchase sites without planning permission or even before they have a clear plan in mind. Nevertheless, the relationship described here between the cost of construction and the cost of sites for building purposes in general holds. Of course, factors other than construction costs affect land prices. Indeed, land speculation depends on anticipating the market price for land, and this in turn depends on a number of variables including confidence in continuing economic growth, government policies and changes in taxation.

It is therefore the owners of land that is capable of being developed who stand to gain the largest benefit of construction cost reductions, not the builders, developers, the eventual building owners or even the government. In practice and in the short run, cost reductions favour those developers who own land and are therefore able to gain from reduced construction costs. However, even these land-owning developers will still need to pay increased prices for sites in the future as the budgets required to cover construction costs are reduced, enabling land developers to increase their bids for sites in competition with other developers.

Similarly, in the housing market the number of houses built does not respond to higher house prices. Although general economic theory predicts that higher house prices would lead to an increase in the supply of residences, this does not occur in the UK housing market. This is because speculative housebuilders know that when house prices rise, land values also rise, and they must be mindful not to sell their stock of land but to hold onto it as they would need to replace land they sold (complete with houses), only to find they could not afford to replace their land reserves, because the value of land had risen in price in the meantime. It is often in the interest of speculative housebuilders not to build houses and sell their land but to retain their rising land assets. As a result housebuilders often choose to maintain the value of their asset base when land values increase, rather than sell the assets off in the form of completed houses.

Given the nature of the housing market just described, setting construction industry cost targets is almost irrelevant. The housing market as with all property markets is as much a land market as it is a housing market. To understand the behaviour of speculative developers and housebuilders, who are themselves one type of speculative developer, one needs to understand the land market. As far as firms are concerned, they are primarily legally responsible to their shareholders to generate profits and protect share value. Any other aims are subordinate to these primary company priorities as far as company directors are concerned. Private sector priorities therefore differ from public sector interests. Proposals to improve the construction industry need to take account of the *causes* and the *context* of problems in order to find practical proposals to deal with them.

Priorities for UK construction

In several reports and strategy plans governments have used target setting in construction as a way of focusing attention on government priorities for the industry. They have done this in spite of the fact that the industry is highly fragmented in terms of the number of firms, most of which are small and medium sized enterprises (SMEs), in terms of the number of trades and skills, and in terms of the variety of work construction firms undertake, ranging from building work to civil engineering, from new build to repair and maintenance. Each type of building work has its own problems and issues. Setting targets has drawn attention away from considering priorities and made the targets ends in themselves.

The question is: what should be the priorities of the construction industry? This is inevitably a question with a political answer, and consequently agreement and consensus cannot be easily arrived at. Nevertheless, it is possible to define a range of possibilities for the construction industry, some of which favour one particular set of interests, while other feasible solutions meet the aspirations of other points of view. Setting priorities rather than targets shifts the debate about improving the construction industry towards discussing competing objectives of the construction sector of different interests and practical suggestions for achieving them. Of course, there will be debate, and different points of view will tend to emerge that propose different practical solutions that favour some at the expense of others. The following proposals are only the opening remarks in a process of reaching a consensus on what the construction industry might look like. Together the priorities suggested in the following chapters form an agenda for purposeful and practical discussion. The difference between the behaviour of the industry compared to these priorities may be used to serve as a measure of the industry's performance.

The proposals in the following chapters may be criticised as subjective and not equally applicable to all firms. They may even be called arbitrary. Nevertheless, the criteria of success in these chapters or something resembling these criteria need to become embedded in construction industry culture over time if we are to have a civilised construction industry fit to serve a modern economy with due regard for those who purchase its output, those who work in it and those who are impacted by it economically, socially and environmentally.

Bibliography

Cabinet Office, (2011) *Government Construction Strategy*, London, The Cabinet Office

Cabinet Office, (2012) *Government Construction Strategy Implementation Report, Government Construction Strategy, One Year On Report and Action Plan Update, Cabinet Office* www.gov.uk/government/uploads/system/uploads/attachment_data/file/61151/GCS-One-Year-On-Report-and-Action-Plan-Update-FINAL_0.pdf

Cahill, D. and Puybaraud, M.C., (2003) "Constructing the Team: The Latham Report (1994)," in *Construction Reports 1944–1998*, eds M. Murray and D. Langford, Oxford, Blackwell

Connaughton, J. and Mbugua, L.A., (2003) "Faster Building for Industry: NEDO (1983)," in *Construction Reports 1944–1998*, eds M. Murray and D. Langford, Oxford, Blackwell

Designing Buildings Wiki, (2017) *Construction Industry Reports*, www.designing buildings.co.uk/wiki/Construction_industry_reports

Egan, J., (2001) "Egan on Egan," *Building*, Vol. CCLXVI No. 8197, 10 August, pp. 18–19

Einstein, A., (1917) "Zur Quantentheorie der Strahlung" [On the Quantum Theory of Radiation] *Physikalische Zeitschrift*, Vol. 6, Leipzig, S. Hirzel www.deepdyve.com/lp/springer-journals/physikalische-zeitschrift-heft-6-1917-7hi0IF0Gy9

Emmerson, Sir H., (1962) *Survey of Problems Before the Construction Industries: Ministry of Works*, London, HMSO

Fairclough, Sir J., (2002) *Rethinking Construction Innovation and Research*, London, Department for Transport, Local Government and the Regions

Farmer, M., (2016) *The Farmer Review of the UK Construction Labour Model: Modernise or Die, Time to Decide the Industry's Future*, London, The Construction Leadership Council

King, V., (1996) "Construction the Team: A.U.S. Perspective," in *CIB Working Commission 65 Organisation and Management of Construction Symposium, Conseil International du Bâtiment*, ed D. Langford, London, E & FN Spon

Laing O'Rourke, (2017) www.laingorourke.com/engineering-the-future/engineering-excellence-group.aspx

Male, S., (2003) "Faster Building for Commerce: NEDO (1988)," in *Construction Reports 1944–1998*, eds M. Murray and D. Langford, Oxford, Blackwell

Moodley, K. and Preece, C., (2003) "Survey of Problems Before the Construction Industry: A Report Prepared by Sir Harold Emmerson (1962)," in *Construction Reports 1944–1998*, eds M. Murray and D. Langford, Oxford, Blackwell

Murray, M., (2003) "Rethinking Construction: The Egan Report (1998)," in *Construction Reports 1944–1998*, eds M. Murray and D. Langford, Oxford, Blackwell

Simon, S.E., (1944) *The Placing and Management of Building Contracts: Report of the Central Council for Works and Buildings*, London, HM Stationery Office

Wolstenholme, A., (2009) *Report Never Waste a Good Crisis: A Review of Progress Since Rethinking Construction and Thoughts for Our Future*, London, Constructing Excellence

3 Replacing target setting

Introduction

Since the Latham Report in 1994, targets have been used in reports on the construction industry in an attempt to galvanise firms to improve their performance and the performance of the industry. However, the target setting approach applied to the construction industry as a whole was only a distraction. Having construction industry-wide targets does not serve to motivate individual firms. Firms have to be able to see short term and longer term advantages for themselves, which invariably come from invitations to tender, a full order book and profit margins. It is up to policy makers to manipulate legislation and fiscal policy to facilitate desirable outcomes by introducing the incentives that drive the behaviour of construction firms.

Game theory has shown that firms adapt their behaviour in the market to optimise their outcomes to meet their needs and priorities given that other parties will be doing the same. The way firms behave is constrained by the rules and regulations set by legislation, which is the government's prerogative to decide and determine. It is then up to firms in the industry to react and respond given their commercial obligations and business skills and their aims and ambitions.

As a starting point for a construction industry strategy, firms in the construction industry need to be focused on producing a quality built environment, improving productivity through training and investment and above all providing the built environment that society and the economy require. Others would argue that these desirable outcomes can appear only indirectly, when firms are engaged in producing in competition with other contractors, but are not the direct result of decisions firms make. This also assumes a belief in the market's ability to deliver goods and services efficiently. However, it is also possible to achieve a quality built environment through state intervention or at least a combination of private enterprise and public sector involvement, as can be seen in the Olympic facilities for London 2012, the road bridges over the Severn, Humber and Forth and numerous other infrastructure projects.

Construction targets are no longer fit for purpose

While targets may be attractive and practical as a management tool at the company level, the question is: how effective are they in motivating the firms that make up the industry? Who owns the targets? Who agrees them? Who takes responsibility for them? Targets within a firm may help to co-ordinate the effort of those working together within particular departments, and they may assist employees working together to produce a specific product or service, but there is no evidence they change working practices when applied to a whole industry.

Within individual firms, where a hierarchy of control exists, targets may well serve as a management tool to direct and monitor production, but industry-wide targets are a very different matter. For one thing, who takes overall responsibility on behalf of an industry to see that targets are met? Within a single firm the responsibility lies with the managing director or their appointee. Across a whole industry there is no-one to take on the role, as it would imply responsibility without authority or power and only a modicum of influence, which could be ignored without penalty.

In their paper on "stretch targets" Santos, Powell and Formoso (2000) argued that targets could in principle be used within construction companies to challenge employees to improve performance and become innovative, but they found no evidence other than finding six firms in the UK and Brazil to support this view.

Meanwhile, Brunet and New (2003) refer to *Kaizen*, the Japanese notion of continuous improvement, which, they argue, motivates workers not through contractual arrangements but more by making use of the employees' desire to contribute to their companies' success without any targets being set. An analogous process similar to *Kaizen* takes place in construction in the UK as firms adopt and adapt to new technologies and materials that arrive on site without the help of contracts but that nevertheless require the skills and goodwill of both the firms and site labour involved.

As Philipe (2015) argues, in the light of the many errors of government-led industrial policy, it might be preferable if governments sought to intervene less in industrial development. However, he goes on to say, a *laissez faire* approach did not always work in the light of well documented market failures. As a result he concludes that governments intervene in markets for a variety of reasons and often with limited success.

Industrial development is important for the growth of industrial capabilities and capacity. It encourages and facilitates the use of new technologies and new industries, leading Philipe to conclude that "economic development requires a mix of market forces and public sector support" (Philipe, 2015, p. 4). This is vital as important sectors of the economy require large investment even before commercial viability has had a chance to materialise. In terms of modern industrial policy, a conceivable purpose of government intervention might be to restructure the economy in the direction of those sectors that represent the most productive and dynamic industries, whatever they may be, whether in manufacturing or services and even in construction, as measured by profitability, growth and potential. Whichever policies are adopted there is a continuing need for the construction industry to remain at the forefront of innovation to facilitate potential productivity improvements and encourage new industrial policies in other sectors of the economy. In this context, construction is central even if it is not seen as highly productive compared to other sectors. It is still necessary to enable the other sectors to function and flourish.

Nevertheless, without any clear vision for the economy as a whole, the targets set in the government's report *Construction 2025* (2013) included lowering construction costs by 33 per cent, speeding up delivery by 50 per cent, reducing greenhouse gas emissions by 50 per cent and increasing exports by 50 per cent. These targets raise more questions: how were the targets selected and set, and by whom and with whose agreement? What were the reasons for choosing these targets without reference to industrial policy in general? The method of reaching these targets is not clear and transparent. The figures appear to be very rounded and arbitrary. Who gains from these targets? And who loses? What if these targets are not met? What punishment would lie in wait for contractors who fail to achieve the targets? And if the targets are met, would that be the end of the story? What would happen next? Could construction firms then take it easy and relax? Are these targets mandatory in any sense or purely voluntary? One needs to ask how these targets were arrived at. How were they calculated? The figures used in the construction industry targets appear to be very rounded and arbitrary.

Construction costs, building prices and land

The government's stated target of construction cost reduction, or more precisely, the reduction in the final value of construction contracts, also ignores one of the most important features of the construction market, namely the role of price in regulating the market by sending out price signals. In any case the role of price in construction differs from its role in most other industries, where any reduction in cost is passed on to customers through the process of competition. In other markets customers respond to lower prices by increasing the quantity they purchase. This can be seen, for example, most clearly in the IT sector, where computers were once expensive and exclusive machines used only by large corporations, government bodies and universities. When the price of information technology declined, sales expanded in the form of personal and tablet computers.

Lower prices do not necessarily lead to increased aggregate demand in construction. Public sector and private sector developers and householders do not in general hire contractors because they are affordable, although affordability is clearly essential. They hire them because they need a new building or structure, not because they are cheap. The cost of construction may appear to be relatively high, and construction projects may be amongst the largest single purchases made by government, consumers and industry. However, because of the durability of buildings and the running costs of maintenance and staffing, the capital cost of a building is not a major obstacle to new build or even undertaking repair and refurbishment. Indeed, the initial cost of constructing a building is relatively insignificant when spread over the whole life cycle of the structure.

Nevertheless, the competition between contractors to build is fierce. It becomes a zero sum game, in which they win contracts at the expense of their rivals. As we have noted, lower prices do not in general necessarily increase the size of construction markets. Instead the benefits of lower construction prices ultimately go to increasing land prices.

In the private sector the value of a building depends only on the rental value or the price the completed building can realise in the property market. Lowering the cost of construction, as the government target suggested, confers little long term benefit to the contractor apart from being able to show a relatively keen price on one project before other contractors copy their new technique, catch up and reduce the cost of construction. The cost of construction may well increase over time, but this is due mainly to the cost of improvements and the raising of standards.

Of course, reaching some of these building cost targets could be achieved by lowering specifications and building inferior structures speedily at the expense of the quality of the built environment, working conditions and wages, and possibly incurring harmful environmental consequences. The top priority is the quality of the built environment rather than the most efficiently built environment. Indeed, these outcomes could be the implied, though unlikely, priorities of government policy. As they stand, the government's 2013 targets were incompatible and inconsistent. They also drew attention away from the way construction projects were delivered and what was actually required of the construction industry in attempting to meet unobtainable, pointless and contradictory targets. For example, speeding up construction tends to raise costs not reduce them, though clearly some techniques may one day make it possible to do both.

Setting industry-wide targets leads to the unjustified impression of construction industry failure and inefficiency. There has to be an alternative to target setting. By setting targets, the industry is skewed in the wrong direction, focusing on the contradictory aims of cost reduction while attempting to speed up delivery and raising output, while increasing exports. Instead, it would be possible for firms in the construction industry to focus on producing a quality built environment, improving productivity through training and investment, improving working conditions and terms of employment and above all providing the built environment that society and the economy require. Alternative priorities might include adequate housing and infrastructure along the lines of the Garden City Movement begun at the end of the nineteenth century by Sir Ebenezer Howard in the UK or the Modernist movement initiated by Le Corbusier with his *Ville Contemporaine* in 1922 in France. An alternative strategy would take into account the wider aims of producing a built environment rather than simply correcting the errors of approach caused by narrow construction production target setting.

Society needs adequate and sufficient housing, factories, offices and infrastructure that enable the rest of the economy and society to function efficiently and sustainably. Even if the industry itself is not as efficient as theoretical economics textbooks might declare, the cost of providing what the rest of the economy needs in the way it wishes the built environment to be delivered may not be captured by the data gathered on construction output, labour or contractors. This may not be in the narrow interests of the construction sector, but it may be a price worth paying for the sake of the economy as a whole. For example, constructing a specific building may not represent the optimum output for the contractor, but it may be the optimum size as far as the client is concerned. In other words the apparent inefficiency of the construction sector may be the price of producing the built environment that enables the rest of the economy to function efficiently.

Some of the targets in *Construction 2025* are contradictory. For example, while one target calls for firms to build much faster, another requires costs to be greatly reduced, presumably without a decline in standards and quality. In reality, while there may be exceptions to the rule, speeding up construction invariably incurs additional costs. Not achieving all of the targets (including the targets set by others, such as Latham (1994) and Egan (1998) in their reports) can only give the unjustified impression of construction industry failure, negativity and incompetence.

An alternative approach to improving the construction sector is urgently required. Such an approach could replace targets with priorities and suggest the means of achieving them. Nevertheless *Construction 2025* contained many potentially useful ideas, but it did not take into account the fundamental economic features and issues of the construction industry that form the source of the problems facing construction.

A report of the House of Lords entitled "Building Better Places" (The House of Lords Select Committee on National Policy for the Built Environment, 2016) was one of the first to replace the central position given to targets that appeared in the majority of reports since the Latham Report of 1994. Instead, this 2016 House of Lords report contains several practical proposals, including encouraging long term planning, increased housing provision and the creation of the post of Chief Built Environment Adviser, similar to the post of Chief Construction Adviser that had existed between 2009 and 2015. Nevertheless, in the 2016 report there was little discussion about the economics of the industry and the reasoning that underpinned the proposals.

The following chapters attempt to explain the behaviour of firms and the economics of the industry, including property development, before considering the priorities of the construction industry. The final chapters use economic theory to suggest proposals to resolve some of the difficulties facing the construction sector and to determine how these solutions might be implemented.

Bibliography

Brunet, A.P., and New, S., (2003) "Kaizen in Japan: An Empirical Study," *International Journal of Operations and Production Management*, Vol. 23 No. 12, pp. 1426–1446

Cabinet Office, (2013) *Construction 2025*, London, HMSO

Egan, J., (1998) *Rethinking Construction: Report of the Construction Task Force*, London, DETR

Latham, M., (1994) *Constructing the Team – Final Report of the Government/Industry Review of Procurement and Contractual Arrangements in the UK Construction Industry*, The Department of the Environment, London, HMSO

Philipe, J., (2015) "Modern Industrial Policy," in *Development and Modern Industrial Policy in Practice: Issues and Country Experiences*, ed J. Felipe, Cheltenham, Asian Development Bank and Edward Elgar

Santos A., Powell, J.A., and Formoso, C.T., (2000) "Setting Stretch Targets for Driving Continuous Improvement in Construction: Analysis of Brazilian and UK Practices" *Work Study*, Vol. 49 No. 2, pp. 50–58

Select Committee on National Policy for the Built Environment, (2016) *Building Better Places*, London, The House of Lords

The House of Lords Select Committee on National Policy for the Built Environment, (2016) *Building Better Places*, House of Lords Report of Session 2015–16, HL Paper 100, Published by the Authority of the House of Lords London, The Stationery Office Limited

4 Setting out the priorities of the broad objectives

Introduction

To find remedies for the problems facing the construction industry, it is essential to know what is causing each problem in the first place. The proposed remedies, taken together, comprise an alternative strategy. The need to replace arbitrary targets with the need to set priorities for the construction industry shifts the debate on construction strategy towards agreeing objectives and priorities for the construction industry and the practical means of achieving them.

Admittedly, the following objectives may be seen as subjective. Some critics may rightly call the priorities arbitrary, and others may see them as not equally applicable to all firms. A suggested list of priorities could be embedded in construction industry culture over time. For example, the following six broad operational objectives could form the core priorities of a proposed vision of an alternative construction industry strategy. Together their purpose is to ensure the construction industry is:

1 a competitive industry
2 an industry that is productive and embraces innovation
3 an industry that produces a quality output
4 an industry that is efficient and whose output is efficient
5 an industry that employs a workforce that is professional in its attitude, behaviour and skills
6 an industry that has an excellent reputation and has confidence and pride in itself

The first step in any plan of action is to state the objectives clearly. These set out the destination and purpose of what is sought. If the objectives are not clear, there is no way of coordinating the actions and behaviour of the many different agents and players in the construction process to achieve a given end. All need to understand the purpose of what is intended in order to collaborate, even while serving their own narrow interests. Based on broad agreement over the kind of building industry the country needs, one of the most important tasks of industrial policy for the construction sector is to set priorities and rewards for the industry that all firms and participants can be incentivised to achieve.

It is vital to appreciate that vested interests and the many diverse groups within construction each have their own sets of priorities, all of which need to be respected. For this reason architects, civil engineers, quantity survey-ors, building firms and trade unions each have their own respective institu-tions that represent their members' interests, even on occasion at the expense of other members of the construction team. As a result they have to represent their interests through formal agreements and collective action. It is rarely possible to suggest an economic solution to any issue that does not ben-efit some at the expense of others. This is the dilemma suggested by Pareto Optimality, which argues that because it is never possible to be certain that the gains to the winners are greater than the losses to the losers, proposals can go ahead, according to Pareto, only if there are no losers, only gain-ers. No solution can go ahead if there are any losers. This does not rule out compensation.

Nevertheless, at the risk of placing Pareto Optimality to one side for the time being, six broad priorities or objectives of the industry are proposed here. In terms of being the starting point of a logical approach to setting con-struction industry objectives they may be seen as general strategic industrial axioms that could be made to apply to a number of other industries and not only construction. The following is a list of axiomatic propositions for the construction industry to perform:

1   Firms in the construction industry should produce a quality built output that is efficient and environmentally sustainable over its life time.

A key priority of the industry is the production of a quality output. This can be measured in terms of the satisfaction expressed by its clients, while meeting their requirements, including sound structures, finishes and func-tionality, while taking into account environmental considerations. Com-plaints and disputes need to be monitored and recorded at the industry level. There should also be a culture of post-construction phase service. This complete service should be balanced with awards and recognition, where praise is merited. An annual report and a review of measures of satisfaction should also be published by a new Office or Ministry for Construction.

2   Firms in the construction industry should form an internationally competitive industry, enabling the long term survival of the skills and organisations that comprise its contribution to the national economy.

Fiscal incentives are needed to ensure the construction industry becomes internationally competitive. International competitiveness demonstrates the relative merits and strengths of construction. However, it is quite possible for the exchange rate to render the domestic industry at a disadvantage in international markets. Nevertheless, it is worth benchmarking the performance of the domestic industry in comparison to other markets to ensure the industry is serving its customers as well as might reasonably be expected. Exports need to be supported by government with tax incentives to ensure firms are not at a competitive disadvantage when trading with any particular country. The construction industry should be seen as competitive, as measured by the international sales of UK-based firms, regardless of their country of ownership or registration. Only by being able to compete with firms in other countries can the quality and value of UK construction be seen as keeping up with the highest international standards. With this in mind, if exchange rates climb beyond the level of competitiveness of UK construction firms, then that would be seen as a natural consequence of a high exchange rate and a prosperous economy.

To foster a climate of improvement and innovation a regular review should be published that monitors competitiveness and gives an assessment of exports and import penetration. Such a publication should complement *Construction Manager*, published by the Chartered Institute of Building (CIOB), and the regular publications of *Construction News* and *Building Magazine*. Other sources of information come in the form of academic journals such as *Construction Management and Economics* and *Building Research and Information*, and these should be seen as sources of information.

3   Firms in the construction industry should be efficient and profitable, enabling the long term survival of the firms that are the vehicle for the delivery of the built environment to the national economy.

The aim of construction firms and professionals should not only produce what clients require but should seek to surpass these requirements and their expectations, and this can be achieved through efficiencies. Efficiency is achieved when the cost per unit of a given output is at its lowest. Construction firms and individuals should be encouraged with awards and recognition, as indeed they are, for example, by the Chartered Institute of Building's Construction Manager of the Year Awards and the Royal Institute of

British Architects' Annual Awards. Moreover, the built product itself needs to meet sustainability criteria. An annual survey of the performance of the built environment and the performance of the industry, including productivity, should be published by the Office for Construction.

4   Firms in the construction industry should be productive and embrace innovation.

The industry should be a safe industry for the people who work in it, with a workforce trained not only in health and safety but also in terms of skills. Those who work in the industry should be able to deliver to high standards of workmanship with a pride in their work and a professional attitude towards their co-workers, clients and others. Qualifications, status and recognition need be reflected in pay scales and terms and conditions. In recent years there has been much discussion amongst industry professionals about the role of image in making the construction industry attractive to new recruits. The image of the industry is important as it is the impression of the industry that attracts new recruits to join the workforce. The better the image of the industry, the greater the number of people willing to work in construction. However, the image of the industry cannot improve until the people who work in the industry are genuinely respected and what they build is appreciated by those within the construction sector itself.

The output of the construction industry is an important component of the productive capacity of the country. Without offices and factories, roads and rail facilities, information technology, energy and water, modern production cannot be internationally competitive. As the Confederation of British Industry (CBI) (2013) recognised in its submission to the government's strategy, construction is an enabler of society and the economy.

The contribution of the construction industry towards the productivity of the rest of the economy is in two forms. Firstly, the buildings have to be produced efficiently, and secondly, they have to perform efficiently. Construction efficiency is not measured by finding the lowest cost while maintaining built quality. This focuses on cost reduction at the expense of quality and service. A satisfaction scale needs to be devised to measure the degree to which the contracting team has satisfied the client. The efficiency of the buildings themselves can be understood in terms of the ratio of new built output to economic growth, a measure that depends on the use building owners and users make of the new built stock and that is not solely dependent on the buildings themselves.

In order to make its contribution to the economy, construction itself must have a labour force whose productivity is comparable, though not necessarily equal, to that in other industries. Where it does not match other industries, the reasons for the differences need to be understood. For example, in repair and maintenance, good and valid reasons for low performance must also be clearly conveyed to policy makers. There are often perfectly valid reasons for construction productivity to fall below that of other industries. For example, poor weather conditions, unexpected ground conditions, late delivery of prefabricated components and late delivery of drawings and instructions can all cause delays and reduce productivity on site through no fault of the contractors concerned. Nevertheless, productivity is ultimately the source of wages, profits and the standard of living of all those engaged in construction. Labour productivity is improved the more plant and equipment are used, and there needs to be a drive to encourage investment in plant and equipment in construction.

5   Firms in the construction industry should be safe and employ a workforce that is professional in its behaviour, attitude and skills.

There is a need to alter the culture of the construction industry workforce. As many people spend their working life in the construction industry, it is important for them to gain a worthwhile life time experience participating in this important economic activity. Since the demise of the craft trades in the late eighteenth and early nineteenth centuries, the pride and status of craftspeople at the beginning of the twenty-first century are in need of reinstatement. At a time when skills are being altered and new techniques are likely to be used in the course of construction, it is important that the quality of work and the motivation of engineers and technicians are not lost. New skills are required, and old skills are still needed for restoration, maintenance, repair, refurbishment and renovation work. The challenge is to motivate the workforce to maintain or improve on the quality of work in the built environment.

6   Finally, firms in the construction industry should have a reputation such that all those who work in the industry have confidence and pride in what they do.

As stated earlier, the image of the construction industry is important as it is the impression of the construction industry that attracts new recruits to join its workforce: the better the image, the greater the number of people willing to work in construction. Nevertheless, image is not as important as reputation. The priority should be to create an excellent reputation for the firms in the industry through the works that they actually produce, regardless of whether they are routine works and projects or iconic buildings and structures. The London skyline and the regeneration of several major UK cities have all demonstrated that it is possible to produce buildings and structures that are admired the world over. These buildings and structures have contributed to the reputation of the UK construction industry, and this in turn has helped to create confidence in the ability of the UK construction industry to deliver on time, build reliably and go on to export some of its services. An annual survey and report of outstanding construction achievements should be widely published and celebrated in the construction press and beyond but with a critique an open analysis of any serious failures, where these may have occurred.

The reputation of the building industry needs to be taken extremely seriously and rests on its output of new buildings and the repair and maintenance of the public realm. Together these projects have contributed to urban regeneration and the reputation of the UK construction industry, while contributing to the culture of the country through its museums, art galleries, theatres, football stadiums, the repair and preservation of ancient and historical structures like Stonehenge, Dundee's new Victoria and Albert Museum and the Houses of Parliament. This in turn has helped to create confidence in the ability of the UK construction industry to build to a high standard reliably and go on to export its services.

One of the stated objectives in the government's *Construction 2025* strategy was to improve the image of the construction industry. Unfortunately, the term "image" implies appearance rather than substance. Reputation, on the other hand, is based on actual performance and is highly valued and easily recognised by firms in the construction industry and their clients. The term "image" should be replaced with a renewed emphasis on *reputation*. Reputation depends on the quality of delivery, and the quality of delivery depends on the quality of training. Reputation is a generally agreed perception that has to be earned over time by delivering beyond expectations.

These priorities together suggest an agenda for a new and fresh debate on construction. Assuming agreement can be reached on these priorities, the debate should then focus on how they may best be achieved. To achieve these broad objectives, several measures are suggested, which together may be used to transform the culture of the construction sector. The following proposals are devised to instil a new attitude at every level in the building industry:

1   In order to measure the contribution of construction to the wider economy a new formal redefinition of the construction sector to include architects, consultants, contractors and others in the supply chain is necessary.

The current data on the construction industry is designed to measure the value added by contractors and subcontractors on site to the inputs of materials, building components and construction services. Therefore, the value of materials, construction components and construction services is not included within the definition of construction and is not included in the measurement of the construction sector in the statistics. Nevertheless, when policy makers, commentators and others discuss the size and relative importance of the construction sector, there is little mention of the fact that in essence all that is being measured is the value of the actual work undertaken by the contractors on site, excluding the value of the materials and building components. Using value added as a measure of construction omits important changes in material inputs. The construction industry is dynamic, and the support, encouragement and continuity of current changes taking place in construction are needed to motivate firms to modernise. Examples of modernisation in construction include continuous technical innovation, project bank accounts and the evolving development of IT. New measures are needed to incentivise firms to encourage innovation and investment with a view to improving both productivity and health and safety in construction.

2   A new approach is needed towards professionalisation of construction skills within an education and training framework for construction qualifications. Skilled craftspeople who work in the construction industry often take a great deal of pride in what they do. This is not always appreciated by site managers or the firms they work in. As a result their work and knowledge are often taken for granted, and the skill, care and effort put into carrying out the work are often overlooked. Raising the level of appreciation of the workmanship involved would do much to motivate the people carrying out the work. This could be achieved if apprentices could enhance their training after completing their apprenticeships and gain professional status through membership in a formal national institute to recognise their skills, experience and qualifications. The professional status should then be recognised through enhanced wages and preferential treatment regarding continuity of work and job security.

3   The development of a national and regional sustainable built environment plan and strategy including the provision of facilities to achieve this is required.

4   To improve working terms and conditions in construction, ensuring the continuity of work as in the infrastructure pipeline with a five year or preferably ten year horizon will enhance job security for the labour force and attract people into the industry. By improving the terms and conditions of employment high calibre individuals would be attracted to take up skilled work in construction. There is no good reason for

people to reject work in the construction industry if it genuinely provides satisfying conditions and is interesting and challenging.

5   There is a need to reform or replace planning regulations and form an integrated housing, health and education facilities strategy – with a minimum five to ten year horizon.

6   By strengthening London as one of the major international construction markets involving contractors, consultants, the English legal system and construction finance, London is set to remain a major centre and market for construction services. One way of achieving this would be to define London as a construction marketplace. This would involve creating a number of organisations to measure and record the construction activities taking place.

At present no record is kept of how much construction legal work is undertaken in London related to national and international dispute resolution, contractual arrangements and construction insurance claims. Similarly, there is no register of construction contracts signed in London to indicate the scale of work being done in the city. Legal services, banking services, consultancy services and design services all play a large and important role, but little if any is recorded as belonging to London in its role serving as a global construction hub.

Bibliography

CBI, (2013) *Building Britain's Future An Industrial Strategy for Construction, The CBI's Submission to the Construction Industrial Strategy*, London, CBI

HM Government, (2013) *Construction 2025: Industrial Strategy: Government and Industry in Partnership*, London, Department for Business, Innovation and Skills, p. 20

5 The size and scope of the construction sector

Statistics on the construction industry have been collected since the Second World War. Although there are sound reasons for the data being collected in the form it is, it understates the size of the construction sector by a large amount. The data is used to calculate the national income by taking the value added by each sector of the whole economy in order to avoid double counting. The result is a consistent total of every part of the economy, avoiding the double counting of the inputs from other industries. For example, the rubber in tires is included with the rubber industry and not counted twice under automotive. This means only the value added by contractors is included under construction, not the materials and components that are brought onto site.

The aggregate of every part of the economy is then used to calculate gross domestic product (GDP) or national income. While this approach might measure the contribution of every sector of the economy, by breaking construction up into its component parts, only that component part that measures construction activity on site is used to measure construction. This measure omits the supply chain and the professional inputs such as architectural and engineering design, manufactured building materials and other elements.

As a result, the data on the construction industry undervalues and understates the scale of this important sector of the economy. However, in the Construction Sector Deal (BEIS, 2018, p. 6), a more accurate presentation of the data on the construction industry stated, "The construction sector, encompassing contracting, product manufacturing and professional services, had a turnover of around £370bn in 2016, adding £138bn in value to the UK economy – nine per cent of the total – and exported over £8bn of products and services." It was usual for the value added figure alone of £138bn to be used to measure the size of the construction industry. It is an improvement in the reporting that the turnover now includes product manufacturing and professional services, measuring the construction industry as contributing £370bn to the economy, not £138bn. A regular publication of these figures would help to demonstrate the importance of construction to the economy.

One of the reasons for studying the economics of the construction industry is to inform government policies on the construction industry to resolve the many issues facing construction, including skills shortages, disputes, lack of productivity in the industry and many more problems.

For example, in 2014 there were 328,000 contractors[1] in the construction sector, of which 300,212 firms employed up to only seven people and 133,737 firms were sole proprietors or firms that employed only one person. Only 688 contractors employed more than 115 people. These firms are divided into a large number of trades under the broad headings of main trades, civil engineering and allied trades in 22 named specialties, such as glazing and electrical installation, and four catch-all categories for the many firms outside any specific trade classification.[2]

As in most industries the construction industry is mainly comprised of small and medium sized firms working in a multitude of diverse trades and transforming a wide variety of materials. As a result of the number of small firms and the variety of trades, the industry itself simply does not have the economic or organisational means to deal with the problems it faces without government intervention. The business environment of the construction industry is in many ways an anarchic situation in which firms are left to their own devices, resulting in expensive litigation and frequent company failures. This is due to the low profit margins of 3 per cent or less of almost all building contractors and the important role played by the vast number of small enterprises in the construction industry, which often act as sub-subcontractors without the full protection of employment law to safeguard their rights and working conditions. Understanding the causes of the difficulties facing the industry is the first stage in suggesting possible solutions.

The construction industry is frequently reported as being only 6–7 per cent of the British economy. For example, in 2012, taking only the value added by the construction industry on site, the industry grew from just more than £116bn in 2012 to approximately £122bn in 2013, a figure equivalent to between 6 and 7 per cent of the UK economy, according to official statistics.[3] As mentioned earlier the size of the industry depends very much on what is included in the definition of the construction industry and what is excluded. For example, construction output data does not take into account the value of prefabricated building components or manufactured products or materials coming from the steel, glass, concrete and timber industries as well as from many other sources.

In 2012, taking the grand total of all the inputs into construction into account, according to Table 5.1 the total value of construction output was more than £208bn (including inputs from other industries). In the same year

Table 5.1 Share of construction industry as a percentage of the national income in 2012

Sector	GVA[a]	Intermediate[a]	Total output (final goods)[a]	Ind TO/ all TO[b]	GVA/total GVA	TO/all GVA[c]
Agriculture	9438	16867	26305	0.94%	0.69%	1.93%
Production	205208	424069	629277	22.50%	15.08%	46.24%
Construction	**86789**	**121430**	**208219**	**7.44%**	**6.38%**	**15.30%**
Distribution	247518	257767	505285	18.06%	18.19%	37.13%
Information and communication	88035	68776	156811	5.61%	6.47%	11.52%
Finance and insurance	116363	117840	234203	8.37%	8.55%	17.21%
Real estate	143641	72287	215928	7.72%	10.55%	15.87%
Professional and support	158811	121575	280386	10.02%	11.67%	20.60%
Government, health and education	258982	208385	467367	16.71%	19.03%	34.34%
Other services	46140	27187	73327	2.62%	3.39%	5.39%
All industries	1360925	1436183	2797108	100.00%	100.00%	205.53%[d]

a £m.

b Industry share as a percentage of total gross output (GVA plus intermediate goods).

c Final goods value to GDP or value of total output (TO) as share of GDP (total gross output less intermediate goods).

d Figure does not sum to 200 due to rounding in the data.

Note: Because total gross output includes intermediate goods, the aggregate total output figure includes double counting. The size of the economy (GDP) is therefore usually measured by deducting intermediate goods. However, the value of any final output includes intermediate goods, and the last column of this table shows the final value of construction over GDP.

Source: Table 2.2, Output and capital formation by industry, gross value added at basic prices, Chapter 2, The Industrial Analyses, National Income Accounts, The Blue Book, 2013, ONS.

the National Income Accounts[4] showed that the GDP of the UK economy amounted to almost £1,361bn, making construction output equivalent to more than 15 per cent of the whole economy, twice the commonly understood value.

Nevertheless, the Office for National Statistics' own data also shows that the combined construction and real estate sectors alone grew from 15.78 per cent of GDP in 2006 to 17.17 per cent in 2013 (ONS, 2015b). This figure does not take construction product manufacturers' output into account, which would add a significant amount of value to construction industry output if it were included. Nor do the figures for construction mention the contribution of architects, surveyors and other consultancies when calculating the size of the construction sector in the economy, although professional services were included in the figures calculated in the Construction Sector Deal in 2018.

This matters as the construction industry is far larger than is commonly perceived to be the case, because official data is based only on the value added on site by main and specialist contractors. What policy makers and decision takers assume they are measuring is often very different from what is actually being measured. While the assumed perception of the size of the construction industry is often based on the value of the completed buildings and structures produced by construction firms, the actual measurement of construction is based on only what contractors do to assemble the materials on site, excluding the materials themselves and what has come to be known as the supply chain.

It is often argued that the construction industry has a multiplier effect, sometimes estimated to be as high as 2.84 (CECA, 2013). According to the Centre for Economic and Business Research referred to in the Civil Engineering Contractors Association (CECA) report of 2013, for every £1bn invested in infrastructure, the GDP increased by £1.299bn. Of the increase in spending 50.5 per cent was due to the construction sector, 8.5 per cent to increased manufacturing and 6.1 per cent to increased wholesale and retail activity. Hence investment in construction, it was claimed, could generate an increase in the economy equivalent to almost 30 per cent of the original investment.

A more reasonable alternative view is that the multiplier effect of the construction industry in the context of an open economy is likely to be of minor importance. Firstly, the multiplier is based on further waves of spending when suppliers themselves become consumers and pass their earnings on to others, who in turn spend their incomes in further rounds of consumption spending. However, in a modern economy the vast bulk of expenditure leaks out of the economy and is rapidly dissipated because little remains in the local economy, leaving the multiplier effect approaching a value of 1. Also the "additional" spending on manufactured construction goods is part of the original investment but is not included in the construction output data and hence exaggerates the impact of construction investment because the value of the investment is undervalued in the first place.

Secondly, the impact of construction investment, especially spending on infrastructure, encourages new developers to invest in additional projects to capture any increase in property values expected as a result of the original investment in infrastructure. This is a completely different mechanism from that described by Keynes as the multiplier effect. And thirdly, the impact of the multiplier assumes the marginal propensity to leak is small enough to contribute to a measurable multiplier effect in a given locality. In fact the multiplier effect would at best be dissipated rapidly and would be very small indeed, especially in the locality of the original development investment.

The economic importance of the building industry can be better assessed as the annual value of the *final* built goods, equivalent to 15.3 per cent of the whole economy, according to Table 5.1. As an alternative measure that takes only the narrow definition of construction, namely the value of builders' output, by 2014 contractors' total output at current prices was £135,146m, of which £84,744m was repair and maintenance. The appropriate construction statistics and data used depend on the question asked. For our purposes it is important to know the size of the industry that best reflects the value of the gross contribution to the economy and also shows the annual (or quarterly) changes in construction.

Notes

1 The number of contractors is derived from the table by summing all the firms in all the trades. The official total is 251,647.
2 ONS, Table 3.4, Number of firms by size and trade of firm, Construction Statistics Branch, Office for National Statistics.
3 Table 13, All work summary, Value of construction output in Great Britain: current prices, non seasonally adjusted – by sector, Output in the Construction Industry, Jun 2014, ONS. www.ons.gov.uk/ons/publications/re-reference-tables. html?edition=tcm%3A77-330935#tab-all-tables
4 Table 2.2, Chapter 2, The Industrial Analyses, National Income Accounts, The Blue Book, 2013, ONS.

Bibliography

CECA, (2013) *Securing our Economy: The Case for Infrastructure, A Report of the Civil Engineering Contractors Association*, May, London, The Civil Engineering Contractors Association

Department of Business Energy and Industrial Strategy, (2018) *The Construction Sector Deal*, London, BEIS, p. 6

Office for National Statistics, (2015a) Construction Statistics Annual, No 16, Table 2.8 Private Contractors: Value of Work Done, by Trade of Firm and Type of Work, ONS www.ons.gov.uk/businessindustryandtrade/constructionindustry/datasets/constructionstatisticsannualtables

Office for National Statistics, (2015b) Nation Income Accounts, Table 2.2 Gross Value Added at Current Basic Prices by Industry, Blue Book www.ons.gov.uk/economy/grossdomesticproductgdp/compendium/unitedkingdomnationalaccountsthebluebook/2015-10-30/chapter02theindustrialanalyses#the-industrial-analysis

6 Innovation in the modern construction industry

Productivity and innovation

Without buildings and infrastructure a modern economy simply could not function nor be internationally competitive. The contribution of the construction industry towards the productivity of the economy takes two forms. Firstly, the buildings have to be produced efficiently, and secondly, they have to perform efficiently. Construction efficiency is measured according to cost of construction, while maintaining the built quality. The efficiency of the buildings themselves is more complex and depends on the contribution of the new built output to economic growth, a measure that depends on the use building owners or users make of the new built stock and that is not dependent on the buildings *per se*.

Construction itself must have a labour force whose productivity is comparable, though not necessarily equal, to that in other industries. This is necessary because the real wages of the workforce depend on its productivity. It is worth pointing out that productivity does not determine the wages paid. The wages paid are determined by the supply and demand for labour. Where productivity does not match other industries, for example, in work on existing buildings and structures, wages will tend to be relatively low. This makes it difficult to attract skilled staff. The only way to pay attractive wages is by setting minimum wage rates. Otherwise, left to market forces alone, wages would fall and surpluses of trained workers would disappear as skilled people leave the construction workforce. This is the reason for the underlying persistent shortage of skilled labour in construction.

The simple definition of labour productivity is output per person per period of time, for example, per year. An industry's annual productivity can be taken as a crude measure of total output divided by total workforce. When construction productivity falls behind productivity in other industries, there are often perfectly valid reasons to account for the differences.

For example, delays caused by slow client decision making procedures, poor weather conditions, unexpected ground conditions, late delivery of prefabricated components and late delivery of drawings and instructions can cause delays and reduce productivity on site through no fault of the contractors concerned. Nevertheless, construction industry productivity is ultimately the source of wages, profits and the standard of living of the vast majority of those engaged in the building sector.

The cause of delays

While it is true that buildings often take longer to build than expected, construction delays are not the major cause of delays in the building sector: planning is. Nevertheless, the planning system is an essential part of the process, as construction proposals have to come under thorough public scrutiny. Few activities impact on local and national communities more than construction projects, and the planning process reflects democratic values. These democratic principles by their nature take time. However, it is true that the planning system is a major cost and may often be frustrating from the point of view of developers and those in the construction industry. This is a bone of contention for all involved in the process as it is a microcosm of the conflicts that arise when the rights of one party impinge on the rights of another. Indeed, there is an ongoing debate on the speeding up of the planning processes involved. The Department for Communities and Local Government recently introduced its National Planning Policy Framework (2012) and Planning Practice Guidance.

A second source of delay outside the control of contractors and the construction industry is caused by developers themselves, hesitating over the decision to build in an uncertain economic climate. Delays may therefore not only be due to contractors but be caused by other parties for a variety of reasons.

Labour productivity can be improved with the use of plant and equipment. This is known as capital deepening. Capital deepening is the result of investment in plant and machinery. In construction, firms tend to hire plant and machinery and therefore invest in equipment only on an *ad hoc* basis. Nevertheless, measures are needed to encourage firms to invest in their own plant and equipment. This has the advantage of placing the workforce on a learning curve as the workers learn to operate the machinery in house and become ever more adept at using it. The resulting increase in productivity can enable both higher wages and profits to be earned.

Improvements in the likely performance of construction firms and the rate of innovation in construction are difficult to measure directly, but they can be assessed using a proxy or substitute measure. Suitable proxies for measuring improvements in contractors' modernisation could be in terms of annual investment in plant and machinery, employment data, including numbers employed and wage rates, and labour productivity. Fiscal measures, such as government subsidies, need to be taken by government to motivate and encourage firms to invest in their own in-house plant and machinery, using hired equipment only for specific or exceptional needs and requirements. In the process, greater use of in-house plant and machinery would possibly help to increase profit margins and in so doing increase direct employment, improving the terms and conditions of employment and giving workers increased security of employment equivalent to that found in other industries.

Bibliography

CBI, (2013) *Building Britain's Future An Industrial Strategy for Construction, The CBI's Submission to the Construction Industrial Strategy*, London, CBI

National Planning Policy Framework, (2012) *Making the Planning System Work More Efficiently and Effectively*, London, The Department for Communities and Local Government, www.gov.uk/government/policies/making-the-planning-system-work-more-efficiently-and-effectively/supporting-pages/speeding-up-the-planning-process

7 A quality built output by a competitive industry

One of the most important key priorities of any industry is the production of a quality output, however that quality may be defined. In construction, environmental considerations have risen up the political agenda and are now also an international issue. The construction sector of each country has a particular responsibility as the pollution from greenhouse gas emissions spills over into other countries, and buildings and infrastructure are seen as a major source of pollution.

It is often assumed that the customers' priorities are the most important ones to satisfy, perhaps because the customers are driving the agenda in most cases. It is certainly the case in a market economy that the customer has a choice, and this gives the customer power over the whole process. Government also has a role to play as guardian of the common good. In addition, as a major participant in the construction market, government is in a position to influence several aspects of construction, including building standards and the health and safety of the output as well as the workforce itself. Other major areas of government influence include employment, training and qualifications, research and development, procurement and payment procedures and the competitiveness of the industry and the firms within the industry. As a major client of the industry government can play a role in changing the way the industry operates by setting the procurement rules of engagement and defining the priorities to which contractors must adhere on public sector projects.

As environmental considerations are a global issue, the construction sector of each country has a responsibility as the pollution of greenhouse gas (GHG) emissions spills over into other countries. Good practice in one country can be cancelled out by the pollution caused by contractors operating in another. Carbon trading has been suggested as one way of dealing with the problem but the extent to which it can be used effectively in such a fragmented sector as construction is debatable. In any case, construction is seen by many as contributing to GHG emissions. According to the Intergovernmental Panel on Climate Change (IPCC, 2014), while the energy sector directly contributed almost half of all GHG emissions, construction contributed only 3 per cent. When indirect emissions from electricity and heating were included in the pollution of the final users of energy, the share of GHG emissions from the building and property sectors increased to almost 20 per cent globally. Consequently, good practice would imply that firms in the UK construction sector take measures to mitigate the GHG emissions in both new build and repair and maintenance, including retrofitting insulation and glazing on existing stock. Clearly, there is a need to provide firms with an incentive to improve their practice beyond simply providing an awareness of the pollution problem. If a government is serious about dealing with pollution, a system of rewards and penalties needs to be introduced. Clearly there are penalties for pollution, but there is no system of reward to encourage good practice apart from carbon trading.

The same anti-pollution measures need to be monitored in both new build and repair and maintenance work, where work on the existing stock of buildings calls for the same duty of care as new build. One of the obstacles to achieving good environmental practice in construction is that the vast majority of work is carried out by small and medium sized enterprises (SMEs), and they do not have the resources or facilities to take on responsibility for environmentally friendly measures to reduce pollution. Again, approaches need to be found to ensure SMEs become involved with these issues, including sustainability and quality of service.

In order to raise awareness of quality issues, complaints and disputes need to be managed with agreed dispute procedures possibly organised at the industry level. Again, setting out the detailed measures required is beyond the scope of this book. However, these measures should be balanced with awards and recognition, where praise is merited. Without a continuous measure of performance, progress cannot be monitored. An annual report and review of measures of satisfaction could be used to inform debate. Quality standards can be raised through the use of insurance guarantees, industry standards and even legislation.

Raising quality requires investment in human capital. Training and education support quality improvement as it becomes necessary to train workers and managers in the use of the latest technology. Those individuals and firms that make use of the latest technology will be the overall winners, and they can do that only if they and their staff are equipped to make use of it and its potential. More key performance indicators may be needed to monitor quality improvements as new ideas spread throughout the sector.

A competitive industry

The UK industry must aspire to be competitive and be comparable to the construction industries of other countries. Being competitive is a measure of performance and is indicative of how aware the firms in the industry are to changes in other countries, learning from them in the process.

Competition may be defined as several firms vying for sales in a given market. Competition in construction encompasses a continuing ability of firms to attract sales with the use of inventiveness and value added and attractive designs that persuade potential customers or clients to accept premium prices because the perceived value is greater than the price due to the relatively high cost of inputs. In economic theory this relates to the surplus value of the product or service being greater than the marginal cost. In other words, the consumer surplus from the purchaser's point of view should be greater than the price, which makes the purchaser willing to pay a price that reflects the value to the purchaser rather than the cost of the inputs plus a profit to the provider.

The competitiveness of UK-based construction firms may be measured by the annual changes in the value of their international sales, as mentioned in Chapter 4. Only by being able to compete with the finest firms from other countries can the quality and value of UK construction be seen as keeping up with international standards. This measure of achievement can be seen in recent years in the automotive industry, which continues to export a large proportion of car production while importing cars from abroad. Official data is needed to review and assess construction exports and monitor import penetration to measure changes in international competitiveness and performance. As the UK is a relatively high cost economy, UK construction firms need to develop value adding techniques to attract foreign sales and demonstrate their competitiveness as has been achieved by the architecture profession and construction and property consultancies, which is a source of pride for those working in those fields. These measures include improving design, introducing new materials, adding new services and more widely using IT in construction and other techniques to improve the services provided by construction firms.

Bibliography

Intergovernmental Panel on Climate Change, (2014) *Climate Change 2014: Synthesis Report. Contribution of Working Groups I, II and III to the Fifth Assessment Report of the Intergovernmental Panel on Climate Change* [Core Writing Team, R.K. Pachauri and L.A. Meyer (eds.)]. Geneva, Switzerland, IPCC, p. 151

8 Efficiency and professionalism

The survival of firms depends on their ability to compete, and their ability to compete depends on their profitability and efficiency. The building industry as a whole needs to be efficient in the way construction output is delivered.

While profitability is the domain of the individual firm, the building industry and the existing built environment as a whole need to be efficient. Construction firms and professionals not only need to produce what clients require but should seek to surpass their expectations. Incentives, such as the CIOB's Manager of the Year Awards, accompanied by financial awards would also help to recognise their effort and motivate individuals to raise their performance across the industry.

One important aspect of the efficiency of the built product can be measured in terms of sustainability criteria. An annual survey of the performance of the built environment in terms of energy use, building improvements and other developments should be published. Again a number of awards and financial prizes could be used to give recognition and raise the reputation of firms and individuals including formal government awards for firms and individuals. Finally, tax incentive schemes could be introduced to encourage and motivate firms to adopt new methods of working to improve performance.

A professional workforce

The image of the industry cannot improve until the people who work in the industry at all levels are genuinely respected and what they build is appreciated, especially by those within the construction sector itself. The industry should be a safe industry for the people who work in it, with a workforce trained not only in health and safety but also in skills and professionalism.

In his paper on professionalism in construction, Fellows (2003) emphasises culture and ethics and refers to the exercise of a particular body of knowledge. While culture and ethics apply to the concept of professionalism, they are usually associated with the attitudes and behaviour of the traditional construction professions, namely architects, civil and structural engineers, the surveying professions and construction managers. Similarly, artisans, such as electrical engineers and plumbers, also have integrity and utilise their technical knowledge and experience for the benefit of their employers. What is distinct here is the proposal to extend the expectations of professional attitudes to the craft skills and trades people. Class distinctions between the construction professions and the trades are no longer valid. Professionalising the trades and introducing values, beliefs and codes of conduct would shift the culture in construction with the creation of higher standards of behaviour expected from craft and skilled workers. This would have the effect of encouraging a virtuous spiral of increased productivity by continuously aiming to improve performance and finding new ways of working as a source of pride.

Plumbers, carpenters, electricians, plasterers and all the other construction trades and skills should have the same expectations of behaviour placed upon them as the more traditional construction professionals. This becomes increasingly relevant as the technical skills and knowledge bases expand as new methods and abilities are required. These workers should be given the respect they deserve, by recognising that their technical knowledge is vital to carrying out their functions effectively. It is their knowledge that enables them to negotiate their fees or rates of pay, but this must be protected or competition for work rather than the value of their work will determine wage levels.

Another very important point requiring a cultural change in the construction industry is that although there has been an emphasis on the image of the construction industry in recent years, this focus needs to shift towards reputation. Reputation is more important than image because reputation is based on actual delivery, while image is based on bias and impression. Reputation is based on real achievement and actual performance. Both professionalism and reputation are key to motivating the workforce and improving performance and satisfaction.

Those who work in the industry should be able to deliver to high standards of workmanship with a pride in their work and a professional attitude towards their co-workers, clients and others. Qualifications, status and recognition should be reflected in pay scales and in terms, conditions and security of employment. The image of the industry cannot improve until the people who work in the industry are genuinely respected and what they build is appreciated by those within the construction sector. Higher wages in the sector would also lead to capital deepening and improved productivity. High wages lead to capital deepening because firms are forced to invest in plant and machinery in order to raise productivity to enable them to pay the higher wages.

Where there is a shortage of applicants seeking work in construction, the labour market behaves in the same way as any other market. Where there is a skills shortage, higher wages and improved working conditions need to be introduced and maintained in order to attract sufficient workers. Continuity of employment requires close monitoring of worker numbers and the location of fully and semi trained individuals in order to adjust the supply of labour available for public works to maintain steady employment. This can be achieved using on-line construction labour market sites to reduce friction in the labour market and improve communication between workers and potential employers.

Training should be undertaken with the help of established colleges. Where certain skills are persistently in short supply, the grant levy system needs to be reformed or replaced. However, the grant levy scheme has served the purpose of supporting those firms that do undertake training in house. This avenue of entry into the industry needs to be conserved, especially where specific skills are required by specialist firms. A legal requirement to employ only qualified staff on site would have the effect of creating barriers in the labour market but would raise wages, productivity and professionalism.

Where there are sufficient numbers of trained workers and a surplus of people willing to work in construction, wages tend to decline, and periods of work are interwoven with periods of unemployment and redundancy. This is a natural consequence of labour markets being left to market forces. It is therefore necessary for government to intervene to ensure continuity of work for those in the construction industry by using periods of recession to invest in infrastructure and other public works to take up the slack in the market, bearing in mind that the cost to the economy as a whole is zero if the alternative is unemployment.

Bibliography

Fellows, R., (2003) *Professionalism in Construction: Culture and Ethics*, CIB Task Group 23, International Conference, October, Hong Kong, Rotterdam, CIB

9 Construction industry strategies

The overarching vision of these priorities is to achieve a vibrant, competitive, internationally engaged construction sector that meets the needs of both its public and private sector clients in safety with reliability and consistency. However, no industrial strategy can offer a panacea for the market volatility of construction and the very low profit margins of the firms that make up the industry.

From the point of view of individual construction firms, from the largest contractors to the self-employed workers, the purpose of defining the value of the construction industry is to raise business confidence and inspire construction firms to create a successful industry, of which all could be proud.

The construction industry is fragmented and divided, often unable to resolve disputes and collaborate and co-operate. To overcome these difficulties, a number of measures are proposed, including the creation of a new industry-wide umbrella body comprising architects, consultants, contractors, specialist firms and others in the supply chain, including product manufacturers and trade unions. Continuing to develop current changes taking place in construction, due to the use of technology and digital applications, can only help to build a vibrant construction sector and make an important, sustainable and long lasting contribution to the economy and society. The developments include project bank accounts and the use of IT, professionalising construction skills and strengthening London as the major international construction market involving contractors, consultants, the English legal system and construction finance.

Construction industry strategies

Defining the industry's priorities is only the beginning of what should be an industry-wide debate, and not all suggestions can be expected to meet with equal approval. When one is discussing these priorities, other priorities are likely to emerge to adapt to changing circumstances and technologies. The result should be a series of priorities that firms themselves need to recognise as being in their own best interest. Adopting them should therefore be to their own benefit and to the benefit of their clients, providing the motivation to change, modernise and improve all aspects of construction. The focus of this vision of the construction industry is on the construction firms themselves achieving recognition and success rather than focusing on targets for the industry as a whole. This approach accepts that firms behave in their own enlightened self-interest and recognises they need to generate profits and returns in order to survive.

The overall vision is to achieve a vibrant, advanced, competitive, internationally engaged construction sector that meets the needs of government and private sector clients in safety with reliability, humanity and consistency. The purpose of this vision for the construction industry is not to offer a panacea for the volatility of construction, but to raise business confidence and inspire construction firms to create a successful industry, of which all could be proud.

However, demand for construction is not the result of the effort made by contractors to make themselves attractive and reliable. This matters only in the context of competing with other contractors for specific contracts. As far as the industry is concerned, once a decision has been taken to build, demand is effective and put to the industry. This depends on the size and scale of projects and the functions the completed structures or buildings have to meet.

In order to achieve these priorities for the construction industry, a number of measures need to be taken. They require political will, industrial leadership and confidence in the future of the construction industry. The following measures are therefore proposed:

1   Create a new industry-wide umbrella body to include architects, consultants, contractors, specialist firms and others in the supply chain, such as product manufacturers and trade unions.
2   Develop current changes taking place in construction, such as project bank accounts and the use of IT.
3   Professionalise the skills within an education plan for construction qualifications.
4   Develop a national and regional built environment plan or strategy.
5   Ensure continuity of work as in the infrastructure pipeline – on a five year or preferably ten year horizon.
6   Reform the planning system to meet a minimum five year horizon.
7   Enact a housing strategy – with a minimum five year horizon.
8   Strengthen London as one of the major international construction markets involving contractors, consultants, the English legal system and construction finance.

10 The construction market

Workflow in construction does not appear on a smooth conveyor belt. Periods of inactivity are followed by periods of intense collaborative working by staff from a variety of skills in many separate, specialist firms, including the use of a variety of materials, technologies and methods of working. On completion of each project the workforce is no longer required, making it incumbent on their employers to find them new or additional work to follow up. The alternative is to face making people redundant. This feature of construction calls for special measures, including public sector intervention.

There are three strands to effectively manage demand and production in construction. The first is to ensure smooth demand and production for construction firms. The second is a national and regional plan and the third a housing plan.

The proposals discussed in this section are only intended to be indicative as the details are intricate and require the interplay of both political forces and interested parties. Continuity of work is vital for firms to have the confidence to employ labour directly and on a permanent basis, providing firms with the opportunity to undertake training. This can best be achieved with a public sector demand management strategy consisting of published plans, detailing forthcoming work well into the future. This first demand management proposal is already in place to a large extent: the government's own infrastructure pipeline of work is a prototype. However, this pipeline of work must be based on at least an eight year commitment, if not much longer, extending beyond the life of one parliament. Knowing future workloads also strengthens the largest firms, helping them to plan and enabling them to compete in international markets.

The second demand management proposal is the preparation of a consistent and co-ordinated national and regional development plan to form the vision of a built environment plan. Such a framework would make use of the Regional Growth Fund and develop a consistent and co-ordinated approach to infrastructure development over the long term. This is not a new idea, but a coordinating strategy has fallen into disuse in recent years. This approach involving national, regional and local authorities to plan and envisage their construction requirements would ensure continuity and co-ordination of projects and employment. The planning system should also be reformed to facilitate major developments, while preserving established democratic principles, an issue currently under much discussion. The process of gaining or denying permission to build ought to be capable of meeting a five year time horizon.

The third element of construction demand management is a housing strategy. According to Inside Housing (2012), the government has recognised the need for 220,000–290,000 houses per annum. Yet it is clear this is not going to be achieved without direct public sector intervention. A housing plan, like the infrastructure pipeline, would also require at least an eight year horizon to begin to plug the hole in housing supply.

The Department for Communities and Local Government has introduced several well intentioned schemes to help both buyers and housebuilders to ease the housing shortage. However, measures to help first time buyers onto the housing ladder, such as "Help to Buy" policies, including the NewBuy Guarantee scheme, are misguided. Such measures only serve to increase house prices above what they otherwise would be and are therefore inflationary. They burden buyers with ever greater debts and largely benefit those selling residential property. However, those policies designed to encourage housebuilders to increase the supply of new housing, such as the housing investment fund scheme entitled "Get Britain Building" or the "Builders' Finance Fund," are to be encouraged, but at only £500m each, these schemes are unlikely to have much impact on total housing needs, helping perhaps approximately 7,000 purchasers, while making it more difficult for those outside the Help to Buy scheme.

The conundrum of the housing market facing government is that solving the housing shortage reduces house prices, and those who have equity in their homes and property backed mortgages would lose out if house prices declined. This therefore has to be a longer term managed change so that existing home owners could alter their investment strategies to take into account that homes might not offer a speculative gain in the future.

London is a global marketplace for construction. Features of London as a global hub for international construction include finance and the English legal system itself. As London is the location of one of the key international financial centres, world class expertise in construction finance is readily available. These resources are indicative of the fact that the UK has a comparative advantage in this area, which is not fully exploited by government or the UK construction sector.

Money laundering is the process of converting cash of dubious origin into legitimate assets through fraud or deception. It is well known that property transactions are often used as a vehicle for money laundering, damaging the reputation of London as a world financial centre. Nevertheless, in spite of having world class construction and engineering consultants and major international legal firms, London is rarely given sufficient recognition as a global marketplace for construction.

The English legal system itself is very important to London as a global construction market hub. The practice of law in the UK is generally recognised as fair and uncorrupted, though, as stated, money laundering through the property market appears to have increased in recent years. In addition, the resources of many of the largest global civil engineering firms and contractors and their services are to be found in the city. Moreover, the consultancy professions based in and around London generate a critical mass of world-leading, readily available skills prepared to undertake the most challenging building and infrastructure projects anywhere in the world.

These resources are indicative of the fact that the UK has a comparative advantage in this area, making UK-based construction-related firms, other than contractors, highly competitive in these highly profitable activities. For many years London has been a destination for the most highly skilled individuals from around the world, who have been attracted by the opportunities of the London international construction market. Recognising the role of London internationally in this respect strengthens London's role as a global centre for construction specific services. However, this position may be weakened by the decision taken in the Brexit referendum of 2016 to leave the European Union. Only the future can demonstrate if this will indeed be the case.

International construction markets

The combination of world class construction and engineering consultants, major international legal firms and the English legal system itself, construction finance and the presence of many of the largest global civil engineering firms and contractors in London means that the UK has a competitive advantage in this field, which is not fully appreciated or exploited by government or UK contractors.

To strengthen London as an international construction market, the effort made by the UK Trade and Industry Department needs to be reinforced by measuring the size of the market, recognising its strengths and training its workforce in the high level skills needed. These measures would strengthen London's role as a global centre for construction specific services and feed into the UK construction sector by setting international quality standards of building and construction services, and encouraging good practice and improvements in the production of the built environment in the UK.

The establishment of London as an international construction market encourages international sales and together with the creation of a new Ministry of Construction and Works, construction demand management and an emphasis on reputation and training completes the strategy to improve the framework of the construction sector. It is then up to firms to respond to the opportunities this strategy would open up.

Nevertheless, the construction sector in a post-Brexit UK could benefit by setting new international quality standards of building and construction services, while encouraging good practice and improvements. This is all the more important for the UK construction industry in the aftermath of the Grenfell Tower Block disaster, in which 71 people died as a result of a systemic failure of the UK building industry, the details of which have still to be published at the time of going to press. However, if lessons are to be learned from this tragedy, then it may be possible to improve practice in building management and regulation as an example to other countries as well as improving work renovating and repairing existing building stock in the UK. This should be one of the roles of any new construction industry or built environment body.

I have mentioned that the vast majority of firms in the construction industry are small or medium sized enterprises, employing up to 250 people. The same applies to virtually all industries. According to the Office for National Statistics (2015c), in construction in 2014 only 246 firms out of a total of 252,000 firms employed more than 300 people. This fragmentation of construction firms reflects the flexibility and adaptability of the industry to meet the ever changing needs of the market. This is illustrated by the fact that construction firms hire labour and plant and machinery as and when required by hiring subcontractors, who in turn hire sub-subcontractors and so on until the work can be carried out in manageable work packages. This is in contrast to large capital intensive manufacturers, which control the specification of their output in standard products, such as specific models of cars, electronic equipment or other similar products, offered to their clients.

It is also in the nature of the construction industry that the very small average size of firms means that the vast majority of firms do not take part in discussions about the industry, its objectives, problems and possible solutions, leaving the debates and participation in policy discussions to the larger firms. This is also the case in the many trade associations and professional bodies that represent the large variety of interests within the built environment sector, most of which are dominated by the largest firms within their specialty. It is therefore not surprising that little notice is taken by the majority of firms of the various reports on the industry. For this reason an alternative approach is needed if change is ever going to take place in construction in a managed fashion. In other words a degree of state intervention is required to influence the behaviour of firms to pursue specific objectives. To this end, a way of motivating firms to make the desired changes needs to be developed using a system of incentives based on fiscal policy and legislation. Discussion on the future of construction needs to focus on specific areas that affect all firms, both large and small, including:

* the scope of the construction industry
* construction demand management
* construction education and training
* international competitiveness of the construction sector

Determining the scope of the construction industry, the first area of focus, draws attention to one of the features of the building industry, namely its fragmentation. As the construction sector is very fragmented, one of the first measures that needs to be taken is to encourage representative organisations across the sector to re-define the boundaries of the sector and the various parties in construction. Even those professions, trades and industries that may wish to remain outside the scope of the construction sector should nevertheless be considered for inclusion in broad definitions of the producers of the built environment.

Secondly, the management of construction demand must be concerned with matching national built environment requirements with resource capacity to ensure construction demand can be met while moderating cost volatility and building delays. Construction demand management is concerned with managing infrastructure and building demand. These developments are already beginning to take shape. For example, the government has set up an Infrastructure Commission to undertake planning and recommend infrastructure priorities. At the same time, reform of the planning system is part of an ongoing debate on speeding up the planning process. The Department for Communities and Local Government set in motion both a National Policy Planning Framework (2012) and a Planning Practice Guidance. It is still too early to judge its success in making planning decisions more effectively and speedily.

Thirdly, publicly funded construction skills education and training requirements are needed, or they will simply not be provided in sufficient numbers. Public investment in the training of skilled workers in construction is investment in the quality of the labour input, leading to greater productivity, higher incomes and improved building standards. The work of the Construction Industry Training Board (CITB) to assess and cost labour requirements of the industry should inform policy in this area. A professional, skilled, knowledgeable, adaptable, well paid labour force is central to achieving a successful construction industry in terms of the priorities of this proposal, facilitating more rapid innovation in construction, raising construction productivity and enhancing the reputation of the construction industry.

Finally, the international competitiveness of UK construction firms and construction-related activities both at home and abroad is to some extent determined by the sterling exchange rate. The diversity of activities related to the production of the built environment implies that some construction services add more value to their inputs in the course of building and designing than others. Consultancies and design practices, such as architects, are in high value added services, and firms in those areas of activity can compete profitably abroad, even when the cost of hiring is high. In many cases their services are not price sensitive even when the foreign exchange value of their currencies is relatively high. Emphasis has been placed on price and undercutting competitors, and insufficient thought has been given to adding value through design, technology and quality of service. A greater effort needs to be invested in research and in applying new methods of working using artificial intelligence, utilising new ways of procuring projects taking into account the life cycle of buildings and structures, improving logistics, using 3D printing to create building components and utilising off-site manufacturing as well as adopting new methods of working through the use of IT and robotics and asking certain questions early on in projects, such as how a project may be made more efficient and how it may generate added value.

Because the UK economy exports financial services and other high value added services abroad, the trend in recent years has been for the foreign exchange value of the pound to increase to the point where construction companies find it difficult to compete internationally, while the converse is true of firms from abroad, which can take advantage of that same exchange rate to price their output more competitively. Currently this trend of sterling to rise has been dented by the move by the UK to leave the European Union. High exchange rates render markets abroad difficult to penetrate, while domestic markets become harder to defend, bearing in mind that the vast majority of firms in the construction sector are too small to attempt to break into international markets. Any serious attempt to increase sales of construction services abroad must therefore be aimed at the largest UK-based firms in the domestic market.

The need to be internationally competitive forces domestic firms to raise the quality of their output to international standards in order to remain in the market. With the arrival of global construction markets it is essential for construction firms to meet international standards of quality, productivity and delivery if they are to have an international presence.

International competitiveness then becomes a benchmark for judging performance, given that exchange rates influence the flows of trade in construction.

Because of the fragmentation of construction the firms in the industry are not able to organise themselves as a productive, co-operative and profitable industry for everyone involved, including labour, specialist firms and clients of the industry.

As a result, in 2016 the House of Lords Select Committee on National Policy for the Built Environment (2016) proposed the creation of a new Chief Built Environment Adviser with a wider remit than the Government Chief Construction Adviser, a position that ceased to exist in 2015. This new adviser's role could be to bring together the many different specialties, trades and professional bodies within the construction and property sectors, some of which do not necessarily have strong links with other parts of the sector. Construction trade unions should also be recognised as having a role to play in such a body, as the whole process is dependent on the role and contribution of the labour input. A construction body, broadly defined, comprising every element in the production process, based on mutual respect, would be able to co-ordinate the trades and professions of the industry and establish a common interest and identity between contractors and non-contractors, which would encourage the integration of the construction process to the benefit of all.

A new definition of the scope of construction to include architects, consultants, engineers, trade unions, construction product manufacturers and those in the property sector would increase awareness amongst policy makers of the size, significance and importance of construction.

Such a body would also continue to encourage and strengthen the introduction and development of a number of changes currently taking place in the UK construction market. These changes include project bank accounts, the expanded use of collaborative IT packages, such as Building Information Modelling (BIM), and the recently formed Infrastructure Commission.

Bibliography

Inside Housing, (2012) *Planning Minister to Raise Direct Payment Concerns*, 8 October, www.insidehousing.co.uk/tenancies/planning-minister-to-raise-direct-payment-concerns/6524090.article

National Policy Planning Framework, (2012) *Making the Planning System Work More Efficiently and Effectively*, London, The Department for Communities and Local Government www.gov.uk/government/policies/making-the-planning-system-work-more-efficiently-and-effectively/supporting-pages/speeding-up-the-planning-process

Office for National Statistics, (2015c) *Table 3.1 Private Contractors: Number of Firms*, Construction Statistics, No. 16, 2015 Edition, Newport, ONS

The House of Lords Select Committee on National Policy for the Built Environment, (2016) *Building Better Places*, House of Lords Report of Session 2015–16, HL Paper 100, Published by the Authority of the House of Lords London, The Stationery Office Limited

11   The way forward

The overall purpose of proposing priorities for the construction industry is to transform the construction industry over time into a modern, productive and professional sector that meets the needs of the rest of the economy in terms of providing its built environment, while taking economic, social and environmental issues into account. The construction industry should be profitable for its firms while offering a safe and fulfilling life for those who work in it and those who use its output.

If the reputation of the construction industry is enhanced, improving the performance of firms through a trained and educated workforce becomes a possibility. This can be achieved if the output of the industry and the way it is delivered match or surpass expectations and international standards. This vision of the industry would then challenge each and every firm throughout the construction sector and its supply chain and restore pride in being part of the construction process. Each firm needs to find its own unique set of solutions, appropriate for its own needs. Firms can achieve this by putting every aspect of their behaviour on a path of continuous improvement based on a set of emerging priorities for construction.

One of the stated objectives in the government's *Construction 2025* strategy, HM Government (2013), was to improve the image of the construction industry. Unfortunately, the term "image" implies appearance rather than substance. Reputation, on the other hand, is based on actual performance and is highly valued and easily recognised by firms in the construction industry and their clients. The term "image" should be replaced with a renewed emphasis on *reputation*. Reputation is a commonly agreed perception that has to be earned over time by delivering beyond expectations.

Meanwhile, training requirements have been in a constant state of flux as skill needs change over time and require constant updating. Having an up-to-date trained workforce is, as in any industrial sector, therefore a key priority for firms in the construction sector. The problem is that construction firms are notoriously reluctant to fund the training needed. From the perspective of individual firms in construction, there is a constant turnover of staff. There is little loyalty. Training labour who then go on to work for competitor firms is unattractive and appears to be a move designed to subsidise competing firms. It is this fear of handing over trained staff to rival firms that deters contractors from training their workforces. The counter argument is that if employing untrained staff is a better strategy, then it makes no sense to train people at all. This accounts for the relatively low level of investment in training in construction.

The grant levy system has been an attempt to overcome the lack of training of construction workers and the problem of fluidity in the workforce, with workers changing jobs and moving from one firm to another. However, because of the perennial shortage of skilled workers, the current grant levy system is failing to train sufficient numbers of skilled workers to maintain or increase construction capacity. A fully funded training and education plan for a new range of building qualifications up to professional status requires public sector support. Entrants to the industry would then be able to complete a period of training that was genuinely geared to high levels of competence. Indeed the ultimate objective of improved training and education is the creation of a skilled building professional elite.

Underlying reasons for the continuing shortage of skilled people in the construction industry are low pay, job insecurity and poor working conditions in construction compared to other sectors of the economy. By increasing wages and improving terms of employment the construction labour market would be in a position to attract a larger number of able individuals, who would then overcome the shortage. Unfortunately, unless demand for construction is maintained over the longer term, these additional workers would be surplus to requirements in a downturn, and both wages and employment would fall, making the industry once again unattractive for career minded individuals. Hence continuity of work is a necessary prerequisite to developing a professionalised workforce.

Ministry of Construction and Works

There is a case for re-establishing a full government Ministry of Construction. The construction industry is dispersed through the country in every constituency and every local authority. The economic impact of construction and construction projects has political implications not only in terms of the economic stimulus provided during the construction phase but also in terms of the long lasting built assets created by the process, benefitting inhabitants for many years to come. There are impacts in terms of economic uplift for a region, education and health benefits and employment opportunities that are opened up by the development of an area in terms of its built environment.

In order to encompass these benefits in a deliberate policy and plan on an ongoing, consistent and integrated basis, a facilitating Ministry of Construction would be best able to facilitate housing, employment, infrastructure, social and cultural features in an environmentally conscious and democratically representative way. The reinstatement of a dedicated Minister of Construction in a designated Ministry of Construction and Works (MCW) would demonstrate government commitment to modernising this large, diverse, significant and unique sector of the economy. A ministerial appointment would signal the importance of the construction sector to the economy and a commitment towards improving the environment.

By coming under the direct remit of a minister, the MCW would be in a position to bring together the many different specialties, trades and professional bodies within the construction sector, some of which have not necessarily had strong links with all other parts of the construction sector. The ministry would be responsible for overseeing the construction strategy. One role of the Ministry of Construction and Works would be to establish a common interest and identity with non-contractors as well as contractors in the industry and help to integrate the construction process in policy terms.

Together with the establishment of the MCW, a redefinition of the scope of construction to include architects, consultants, engineers, construction product manufacturers and the property sector would increase the relative size, significance and importance of construction in statistical terms, signalling the importance of the sector to the rest of the economy.

The economic features in this summary of the economics of the construction industry are not exhaustive. Many other factors also influence construction. Nevertheless, these features cannot be ignored by any new construction regime. Construction strategy discussion needs to move away from a target setting culture, where no one can be held responsible for failing to adhere to the targets. Instead the strategy needs to be based on continuous improvement and review. Improving the performance of firms through a better trained and educated workforce is a prerequisite for enhancing the reputation of the construction industry, which is achieved when the output of the industry and the way it is delivered match or surpass expectations and international standards. This in turn depends on firms throughout the construction supply chain finding their own solutions to their own technical and managerial problems by moving towards a path of continuous improvement with the support of a national framework for construction managed by a new integrated Ministry of Construction and Works.

Perhaps a suitable vision for the construction sector is for it to become and remain a modern, productive and professional industry that meets the needs of the rest of the economy in terms of providing its built environment, while taking environmental issues into account. Such an industry would provide profit for its firms and a healthy and safe environment for its members and those who use its output, and would offer a safe and fulfilling life for its workers.

Bibliography

HM Government, (2013) *Construction 2025: Industrial Strategy: Government and Industry in Partnership*, London, Department for Business, Innovation and Skills, p. 20

Appendix

List of reports on the UK construction industry, 1934 to 2018

Those titles in bold have been referred to in the text.

1934. Reaching for the Skies. Alfred Bossom.

1944. Simon Report: Placing and Management of Building Contracts.

1948. The Distribution of Building Materials and Components. Sir Ernest Simon.

1948. The Working Party Report to the Minister of Works: The Phillips Report on Building.

1962. Emmerson Report: Survey of Problems before the Construction Industries.

1964. Banwell Report: The Placing and Management of Contracts for Building and Civil Engineering Work.

1965. The Future of Development Plans. Planning Advisory Group.

1966. Tavistock Report: Interdependence and Uncertainty: A Study of the Building Industry.

1967. Potts Report: Action on the Banwell Report: A Survey of the Implementation of the Recommendations of the Committee under the Chairmanship of Sir Harold Banwell on the Placing and Management of Contracts. Economic Development Committee for Building of the National Economic Development Office.

1969. Skeffington Report: People and Planning: Report of the Committee on Public Participation in Planning.

1970. Large Industrial Sites. National Economic Development Council.

1975. Wood Report: The Public Client and the Construction Industries: The Report of the Building and Civil Engineering Economic Development Committees Joint Working Party Studying Public Sector Purchasing.

1978. PIG Report: Project Information – Its Content and Arrangement. A Report and Proposals on the Way Forward, by the Project

Information Group (PIG) of the Department of the Environment NCC Standing Committee on Computing & Data Co-ordination.

1980. Engineering Our Future: Report of the Committee of Inquiry into the Engineering Profession, HMSO. Sir Montague Finniston.

1983. The British Property Federation Manual of the BPF System for Building Design and Construction.

1983. Faster Building for Industry. National Economic Development Office (NEDO).

1988. Faster Building for Commerce. National Economic Development Office (NEDO).

1993. Latham, Trust & Money.

1994. Latham Report: Constructing the Team.

1995. Progress through Partnership: Report from the Steering Group of the Technology Foresight Programme.

1995. Construction Procurement by Government: An Efficiency Unit Scrutiny. Sir Peter Levene.

1996. Educating the Professional Team. Construction Industry Board.

1996. Constructing a Better Image. Construction Industry Board.

1996. Training the Team. Construction Industry Board.

1997. Framework for a National Register for Consultants. Construction Industry Board.

1997. Liability Law and Latent Defects Insurance. Construction Industry Board.

1997. Partnering in the Team: Report of Working Group 12. Construction Industry Board.

1998. Strategic Review of Construction Skills Training. Construction Industry Board.

1998. Egan Report: Rethinking Construction.

1999. Achieving Excellence. Office of Government Commerce.

1999. Towards an Urban Renaissance. Urban Task Force.

1999. A Better Quality of Life – A Strategy for Sustainable Development for the United Kingdom. Government.

2000. Better Public Buildings: A Proud Legacy for the Future. Department for Culture Media and Sport.

2000. Construction Industry Board Root and Branch Review. Construction Industry Board.

2000. A Commitment to People: Our Biggest Asset. A Report from the Movement for Innovation's Working Group on Respect for People. Movement for Innovation.

2000. A Vision Shared: The Movement for Innovation Second Anniversary Report. Movement for Innovation.

2000. The Housing Demonstration Projects Report: Improving through Measurement. The Housing Forum.

2000. Building a Better Quality of Life: A Strategy for More Sustainable Construction. Department of the Environment, Transport and the Regions.

2001. Modernising Construction. National Audit Office.

2002. Rethinking Construction 2002: Achievements, Next Steps, Getting Involved. Rethinking Construction Group Ltd.

2002. Accelerating Change: A Report by the Strategic Forum for Construction. Chaired by Sir John Egan.

2002. Fairclough Report: Rethinking Construction Innovation and Research: A Review of Government Policies and Practices. DTI and DTLR.

2005. Improving Public Services through Better Construction. National Audit Office.

2005. Be Valuable. Constructing Excellence.

2005. The Urban Renaissance Six Years On. Urban Task Force.

2007. Callcutt Review of Housebuilding Delivery. John Callcutt.

2007. Low Carbon Building Standards Strategy for Scotland (the Sullivan Report).

2007. Place-Shaping: A Shared Ambition for the Future of Local Government. Lyons Inquiry into Local Government, Sir Michael Lyons.

2007. The Construction Research Programme – Project Showcase. Department of Trade and Industry.

2008. ICT and Automation (ICTA) Scoping Study Report. National Platform for the Built Environment.

2008. The Strategy for Sustainable Construction. Government/Strategic Forum for Construction.

2008. Construction Matters. Business and Enterprise Select Committee.

2008. Equal Partners. Business Vantage and Construction Clients' Group.

2009. Never Waste a Good Crisis. Andrew Wolstenholme, Constructing Excellence.

2011. Government Construction Strategy 2011. Cabinet Office.

2011. NAO, Lessons from PFI and Other Projects. www.nao.org.uk/report/lessons-from-pfi-and-other-projects/#main-content-anchor.

2011. Seventeenth Report: Private Finance Initiative. House of Commons, Treasury.

2011. Laying the Foundations: A Housing Strategy for England. HM Government.

2011. Portas Review: An Independent Review into the Future of Our High Streets. DCLG and BIS.

2012. A Better Deal for Public Building. Report from the Commission of Inquiry into Achieving Best Value in the Procurement of Construction Work. All Party Group for Excellence in the Built Environment.

2012. Government Construction Strategy: Final Report of the Procurement/Lean Client Task Group. Government Construction Task Group. July 2012.

2013. Construction 2025: Industrial Strategy. Government and Industry in Partnership, London. Cabinet Office.

2013. BIS. Professor John Perkins' Review of Engineering Skills.

2013. Global Construction 2025: A Global Forecast for the Construction Industry to 2025. Global Construction Perspectives and Oxford Economics.

2014. Our Future in Place. The Farrell Review of Architecture + the Built Environment (FAR). Sir Terry Farrell.

2014. Property in Politics. Royal Institution of Chartered Surveyors (RICS).

2014. Lyons Housing Review: Mobilising across the Nation to Build the Homes Our Children Need. 16 October 2014.

2014. Skills to Build. LCCI/KPMG Construction Skills Index (London and the South East) 2014, London Chamber of Commerce and Industry (LCCI). November 2014.

2014. An End to Cold Homes – Labour's Energy Efficiency Green Paper. 10 November 2014.

2014. Going for Growth: Reviewing the Effectiveness of Government Growth Initiatives. All Party Urban Development Group. 17 November 2014.

2015. Collaboration for Change: The Edge Commission Report on the Future of Professionalism. Paul Morrell. 18 May 2015.

2015. Fixing the Foundations: Creating a More Prosperous Nation. 10 July 2015.

2015. A Brighter Future for Our Towns and Cities: A Report from the Commission for Underperforming Towns and Cities. The Institute of Economic Development (IED).

2015. Government Construction Strategy. Cabinet Office.

2016. Government Construction Strategy, 2016–2020. Cabinet Office.

2016. Building Better Places. The Committee on National Policy for the Built Environment. 19 February 2016.

2016. Government Construction Strategy 2016–2020. Infrastructure and Projects Authority. 23 March 2016.

2016. Farmer Review 2016: Modernise or Die.

2016. Bonfield Review: Each Home Counts: Review of Consumer Advice, Protection, Standards and Enforcement for Energy Efficiency and Renewable Energy. Department for Business, Energy & Industrial Strategy, Department for Communities and Local Government.

2017. Building our Industrial Strategy: Green Paper.

2017. Housing White Paper: Fixing Our Broken Housing Market. 7 February 2017.

2018. Construction Strategy: Construction Sector Deal. Department for Business, Energy and Industrial Strategy, BEIS, Open Government Licence, London.

Index